# THE HUNGER
# FOR MORE

# THE HUNGER FOR MORE

*God, Gravity,*
*and the Big Bang*

## PETER STRASSBERG, M.D.

*Full Court Press*
*Englewood Cliffs, New Jersey*

Published in the United States of America
by Full Court Press, 601 Palisade Avenue
Englewood Cliffs, NJ 07632
fullcourtpressnj.com

ISBN 978-1-938812-37-8
Library of Congress Control No. 2014945826

*Editing and Book Design by Barry Sheinkopf*
*for Bookshapers (bookshapers.com)*

*Graphic Art by Richard Donatone*

*Cover Art courtesy Hubble Telescope*

*Colophon by Liz Sedlack*

TO ELAINE

*my constant companion*

# PREFACE

I still remember that somewhat portly, pleasant, elderly man, our family doctor, who would come to our apartment when we were sick. My older sister, and I and my two younger brothers, would try to hide our feverish bodies under blankets, hoping not to be seen. However, we were always uncovered, examined, and routinely given penicillin injections. It didn't matter what we were sick with; we always were given a shot.

My mother then would sit with this physician, Dr. Halbein, and listen in rapt attention to his recommendations. I guess it was that deep, profound respect she showed to this person that led me into my profession. I always wanted to be The Doctor, the one who understood the problem and attempted a reasonable solution.

At this stage of my life I have studied and practiced medicine for over forty years. My reasons for entering and staying in the field have changed as my life situations have changed. In this respect I am like everyone else. As a young man helping to build a family, monetary rewards are of primary concern. However, as one's family matures and becomes wholly independent, incentives begin to change. We realize that things will not continue forever, and, therefore, whatever knowledge we have gleaned should be imparted for other's use. A doctor then, at the most basic level, is a teacher.

This book is a journey. It is an attempt to understand the forces that shape our world. My background as a physician allows for a rudimentary knowledge of different sciences. I do not profess any great, deep scientific background in any of the hard sciences—physics, chemistry, biology—and their use in this book is quite cursory and should be reasonably easy to follow. I do not expect my readers to stumble while trying to understand where I am going.

I use some minimal mathematics. I know that, as soon as one formula is written, half my readers are gone. However, a little math is necessary, as there is no other way to convey several ideas. I promise the math will never be onerous.

Exponents, powers of ten, are used to simplify very large and very small numbers. Thus, when writing 1000 one could write $10^3$, or when writing 1/1000 one could write $10^{-3}$. A commonly used example is the speed of light. It can be written either as 300,000,000 or as $3 \times 10^8$ meters/second. When multiplying, one just adds exponents, and when dividing one subtracts. In general, they make things more concise and should not be a distraction.

The book itself is in the form of a dialog, with questions and answers that I hope make some sense, and that are never meant to be pedantic. This form was inspired by conversations with close friends.

The book is also in memory of a wonderful brother-in-law—Bob—and a loving son-in-law—Mark—whose journeys ended too soon.

Finally, I wish to thank my wife Elaine for both her tolerance and good humor. It was she who kept my head on straight.

—*Peter Strassberg*
*November 2014*

# TABLE OF CONTENTS

## Part One
## THE ESSENTIAL CAUSE: THE HIGHER DIMENSION

@

## Part Two
## THE PRIMARY EFFECT: THE PHYSICAL WORLD

*Part Three*

## REAPPRAISAL

# Part One

## THE ESSENTIAL CAUSE:
## THE HIGHER DIMENSION

*HE FIRST PART OF THE BOOK attempts to show the existence of a higher dimension. It is found in all things—our daily pursuits, our religious yearnings, our society's fundamental interactions. It is perceived as an underlying force that allows us to grow and evolve, to acquire and accumulate both physically and intellectually; many conceive of it as God.*

*Scientists see this higher dimension as they peer ever deeper into the cosmos and mistakenly consider its changes—the continually increasing redshift—to be due to an expanding universe. These findings have then been interpreted as the Big Bang theory, an initial "explosion" from some primeval "atom." However, this theory is shown to be based on a fallacy, as there is no expansion of space; the reality, a higher-dimensional curve, is discussed and shown to be the underlying cause.*

*Thus, the higher dimension, God to those of faith, is the reason for what exists. Given "the essential cause" the latter parts explore "the effects"—the physical world and the human realm.*

# SECTION I

# *Current Concerns*

THE FOLLOWING CHAPTERS DISCUSS the consequences of an ever-increasing abundance as our societies evolve. This continual improvement in the ability to acquire and accumulate is due to an underlying force of attraction that permeates all things; however, it also leads to some fundamental problems—irrational energy absorption, or obesity, and excessive, burdensome free time.

Our attempt to overcome the problem of unneeded time gives rise to multiple fillers or diversions: games, drugs, alcohol, and superfluous, unnecessary work. The most important of these is, however, religion. It is basic to society for it helps us to understand this underlying force; and its incorporation into a society leads to a more successful and coherent one.

In general, accomplished societies acknowledge this underlying force that allows their members to maximally acquire and accumulate; these are free societies. They also have a strong religious underpinning, as this helps to cement all facets. The United States is such a society, as its powerful spiritual base and essential freedom allow for its growth and prosperity.

# 1

# OVERABUNDANCE

*OUR AFFLUENCE, OUR PLENTY, is due to an underlying force that permeates all. In human terms, it is seen as a continual desire to acquire and accumulate—a hunger for more. It leads to a progressive increase in energy and knowledge, but its perverse aspect is the overabundance found in individuals: obesity, with subsequent vascular problems.*

*If we define disease as competition with other energy-absorbing (life) forms, then obesity and its consequences (heart and vascular ailments) are not diseases; they are merely conditions brought on by an irrational, excessive desire of accumulation—eating too much. The problem of vascular "disease," then, becomes secondary to continually evolving society's ever-greater efficiencies.*

## Disease and Competition

Over forty years have passed since I first started training as a physician. Some things have significantly changed, while others have remained constant. Patients no longer view physicians as omnipotent healers but now more realistically question their advice and frequently seek alternatives to their therapies. Mistakes by physicians, real or imagined, are more rigorously challenged, and physicians' costs in defending themselves have soared. Payments to physicians now are invariably other-party dispersed—the government and private insurers being equal contributors.

These are the surface changes driven by society's altering customs and values; but an underlying constant remains, one that predates scientific medicine. Common people with ordinary fears and pains seek advice and comfort from their brethren. Physicians still fill this need and will continue to as long as people live together.

Although medical treatments and diagnoses continue to evolve, what would you say if I told you that a good portion of all medical deaths in the United States were not disease related?

*Question:* Wait a second, you must mean heart disease and its related problems, for I read somewhere that these ailments alone, each year, account for almost one-half of all deaths in the United States. Are you saying that heart disease is not a disease?

*Answer:* That is exactly what I am saying. What we usually consider to be disease is not really so. Much of what we in the Westernized world suffer from is due to our society's efficiency and success. The main problems, heart disorders, hypertension, diabetes, strokes, and other vascular conditions, are really secondary to the overabundance surrounding us. By partaking of this excess, we injure and kill ourselves. So these are not diseases; these problems are simply caused by overly efficient accumulation and irrational ingestion.

*Question:* You are getting too abstruse. What do you really mean?

*Answer:* Well, I am trying to define disease as the consequence of some fierce competition. A bacterial infection, thus, would be a disease (some invading organism attempting to use our energy, our being, for its sustenance). Competition for our essence has always been a cause of misery, and, therefore, throughout history illnesses and death have most usually been associated with infections or starvation.

Larger creatures (wild beasts, mice, or insects, to name but a few) have also frequently been our opponents. In a sense, being attacked and eaten by a fierce animal is closer to a disease than is a heart attack or a stroke. Of course, people have always been their own greatest predators, and war is similar to these diseases. In all these examples, the bottom line is a continual battle for essentials.

## Excessive Accumulation

Heart and other vascular problems are entirely different. In these modern ailments, there is no such rivalry. Instead, an overabundance of energy is secreted away, put into safe hiding in case of future shortage; but the act, the accumulation itself, is the cause of the ailment. Thus, there is no struggle for our energy, just a perceived but irrational need, in the presence of over-abundance, to accumulate, and illness and death follow.

*Question:* So you are defining disease solely as competition. However, aren't you leaving out many important categories? Where would you place cancer, or trauma, or poisoning, or genetic and immune concerns? I could go on and on, but my point is that your definition appears too narrow.

*Answer:* You are entirely correct; I am confining it to competition. Historically, infection (or competition) always was the number-one killer. However, today it has been supplanted by vascular *disease*; thus, my comparison and definition are reasonable, if limited.

*Question:* Okay, to get back to competition, all contests require at least two sides. Thus, what we consider disease would be necessary and appropriate to other species.

*Answer:* Yes, competition means more than one approach. There are always winners and losers; health implies success, disease failure. Competition always was, and always will be, the cause of the natural order of things. However, when societies began to civilize, to grow crops, to herd animals, to accumulate more food than was needed, this innate harmony went awry.

We still have the impulses and desires of primitive, wild animals but now have an abundance of food and energy never before available. As we still are driven by our basic needs we continue to accumulate, to eat as if there will never be more—for there never really was more than on a fleeting basis. But society, that semi-rational grouping of humans for efficient energy utilization, in time created that never-before-existent *more*.

Since there now was abundance overeating, a perfectly effective way of surviving nature's feast and famine has become today's great dilemma. Binging, in times of plenty, is necessary, as a wild animal never knows what, in the future, may be available. Even if not stolen by others, food will certainly rot—be consumed by bacteria or fungi. The only sure way to maintain food is to eat it. We, like all other animals, ably store this digested food as fat, the most efficient form of energy.

Thus, a wild animal eats and eats but rarely gets fat as, in general, there is barely enough—competition sees to that. An excess of food leads to more animals at the feast, or more offspring. Soon the excess is gone, and the ensuing lack can cause severe conflict and death. So there are very few times when a wild animal can be fat or store excessive amounts for long enough periods to lead to significant vascular changes. A tame animal, a civilized human, however, can continue to accumulate, and if the society one lives in is stable, need never face starvation. Therefore, a domesticated individual can over-accumulate and ultimately fill all reserve areas of its body with fat, finally succumbing to this excess.

Thus, civilization, as it has taken a wild animal with natural and untamed desires and placed it in the arena of overabundance, has caused what we mistakenly think of as disease. But this new vascular engorgement, this filling of all our crevices and niches with fat (energy stored for future use), is not a disease, it is a problem based on a misunderstanding. It is the enigma of inadequately comprehending that, once hard natural competition has ceased, equally demanding self-regulation must take its place. We have fooled nature, but we must not fool ourselves. As we no longer are wild animals we should not act as such.

## Dieting—Rational Starvation

*Question:* So, if you are saying that vascular disease is simply a problem based on overabundance, what do you suggest as its solution—starvation and war?

*Answer:* That has always been a *natural* solution but obviously not the most pleasant one. I certainly hope we do not face it. No, I suggest a rational alternative; in fact I wrote a book on this topic—*The One Hundred Year Diet*. Simply, since over-accumulation is the culprit, *sensible* accumulation is the remedy. For those excessively fat or even somewhat obese, dieting—controlled, rational starvation—is the answer. The diet I suggested in my book was a high-protein, high-fiber diet, a natural diet, one associated with our evolution. Anyway, my basic point is that, in an era of overabundance, the reasonable approach is underutilization—eat less!

*Question:* I think I understand. You feel that many of the medical problems of modern Westernized society (vascular complications such as heart attacks, strokes, high blood pressure, and diabetes) are not true diseases but simply misfortunes brought on by efficient social patterns. Our society is so resourceful that more than is needed is found. We, in turn, still react as did our ancient ancestors and we, still driven by our basic animal instincts devour more than is good for us.

This constant overeating is what leads to excessive fat, with subsequent vascular insults. But over-accumulation does not fit your definition of a disease (it is not competition by another energy-absorbing form), it is simply our inability to rationally overcome our basic desires when society is too efficient.

*Answer:* Yes, that is what I mean by it not being a disease. In fact, the more capable one is at absorbing and maintaining energy or fat, the better that person would do in a truly wild state. Obviously, those of us who can survive longest without food would do best during scarcity. Since famine was our ancestors' frequent companion, we are but the product of their survival.

Vascular "disease" is an adaptation to starvation. It is a very positive survival attribute and is very common in our society. It is only in the unusual circumstance in which we find ourselves today (with abundant resources) that it has become a detriment. The ability to absorb and store fat has always been essential to survival. It is not a disease; it is a blessing.

*Question:* So, if it is not a disease, if it is simply our basic desires running amok, why do we allow such self-defeating actions?

*Answer:* We find ourselves (and all other things) driven by an underlying, relentless force. On a cosmic scale we see it as gravity; in human terms we feel an overwhelming desire to acquire, to accumulate—*a constant hunger for more.* This fundamental need controls our every action, and vascular *disease* is but one consequence. We will make an attempt, as our discussion unfolds to understand it; however, for now, we will simply mention that society's ever-increasing efficiency leads to both our success and our failure.

# 2

# CIVILIZATION

*C*IVILIZATION, *WITH ITS PLENTY, due to an ever more efficient acquisition of energy, leads to the problem of spare time. Time not needed to find food or other sustenance must be filled, or it becomes too burdensome.*

## The Curse Of Abundance

If we were to expand somewhat on our previous ideas concerning mankind's evolution (that we have evolved to survive in a difficult world), we will find, in reality, that we have fallen out of our evolutionary niche.

**Question:** Wait a minute. What do you mean? Is evolution not just an ongoing process, an inevitable fact? How can something be out of its evolutionary niche? We are just travelers in a journey that only God knows; how, then, can we be at a wrong place?

**Answer:** Well, you are right. All things obviously are where they are and must survive or perish wherever they find themselves. All animal and plant species are in the same predicament. But no group goes on forever. Some do better than others. Their offspring thrive. Others not as lucky or smart, or as well adapted, do poorly.

We humans, although evolved to a very high level, are essentially out of step. We have evolved so fast that we no longer live as originally designed. We have become so efficient at energy accretion that we now have extra, unneeded time. We have become civilized. This is at the same time our blessing and curse.

**Question**: But how could civilization be a curse? It takes us from a fierce and chaotic world, with scarce food resources, and places us in the somewhat tamer and less violent abundant life that we now enjoy. We certainly live longer by being civilized. We certainly can support a larger population with our greater availability of energy and food. So civilization must be a blessing.

Is it because we now possess terrifying weapons potentially capable of destroying our world? I do not see any other way it could be a curse. If we are lucky or intelligent enough to live in peace or, if at war, nuclear-free, we will not destroy life. Do we not have electricity, do we not have fast and efficient transportation, do we not have adequate sanitation, are we not effectively combating disease, are we not growing more abundant crops? Surely our blessings far outweigh our curse.

**Answer**: You are partially right, yet you miss my point. Let me explain. What you say is correct, for we have become ever more efficient at acquiring energy. This is the natural evolution that all things take. We have accelerated because of our innate mental competence and our ability, within society, to live and work together. We have succeeded beyond our ancestors' wildest dreams. If you count these as blessings, we are surely blessed.

However, we have advanced too quickly. Not because it is bad to have more—no, on the contrary, it is good. Not because it is bad to live longer. No, obviously most of us desire to live as long as is possible. These things in and of themselves are all good.

No, the curse is not in the good that organization and efficiency have brought. The curse is that we have not biologically adapted to fully partake of this good. Our biological evolution is much slower than our societal growth. We are still wild jungle animals. Our drives—our very beings—are adapted to that state; our appetites, our foods and lusts, are those of feral beasts. It takes many, many generations, many more than we can count from the first inkling of civilization, to change these basic biological drives. The curse of civilization is not in the abundance and freedom it allows, it is that we are not able to properly incorporate these blessings.

This inability to change presents us with a fundamental problem: free time. In the wild state an animal has little or no spare time. An animal searches for food, defends what it finds, and hopes for companionship. There is minimal extra time. However, civilization, by domesticating us, altered that.

Humans originally were hunters and gatherers. Prehistoric humans, as we understand them, lived in roving bands of perhaps twenty to thirty individuals. The hardier souls—the mature, younger men—protected the home and searched for food. The women and children—and the older, feebler men—helped by gathering small game and plants.

The community prospered if food was abundant; it suffered if food could not be found. There was little spare time, for there were few occasions when excess food was available. If game was found, it had to be eaten quickly and defended at all times, for other, stronger animals frequently lurked nearby. Even if not stolen, it would, in a brief period, almost certainly rot and be useless. Thus, most time was spent in the planning of a search for, or in the actual gathering of, food.

## Free Time—A Blessing And A Curse

When we became as efficient and organized as to understand that animals need not necessarily be hunted but could be controlled or herded; when we discovered that plants grew from seeds, and could then grow what was needed; we then became civilized. This probably happened 10,000 to 20,000 years ago. However, with it came the new problem: free time.

We now had time we did not need. We had time to ponder, to think. None of us can stop this process. We cannot close our minds as we shut our eyes. Since thoughts continue at all times, we now had time to ourselves just for contemplation. Biologically, we had little use for this, as in previous periods it had only fleetingly occurred. So we had something given to us by our efficient organizing ability that was inherently foreign. How to assimilate this unfamiliar thing, this extra freedom, became the curse that accompanied our blessing of plenty.

*Question:* So you are saying that, no matter how successful we become, this curse will always follow the blessing. This problem of free time, the result of our ability to organize and efficiently reap energy, is inherent to our success.

*Answer:* Yes, free time, time in which there is no significant thing to do, is the challenge that we must deal with, as it has no real parallel in nature. Humans are not designed, any more than are other living things, to deal with useless time. So instead of being pleasant and beneficial, it often becomes burdensome and unsettling.

This dilemma—excessive, unneeded time—probably first surfaced with the abundance that civilization brought. Humans were always social animals. The hunting groups in which people lived tended to have their own territories. They stalked large and small game, and gathered plants and other edibles. Lacking sufficient concealment or the ability to store food for long periods, they had to continually plan new pursuits and new gathering spots. They probably moved with the larger animals from watering spot to watering spot.

So life was set. You hunted, you gathered, you protected, and if successful you celebrated. Then you started again. Little time was left that was not part of your biological program. The wild state's curse was scarcity. But this changed with civilization.

*Question:* It does not seem so pleasant to be in the wild, to live in a cave with inadequate heat, with little protection, with disease and starvation as common boarders. I will gladly give this honor away and accept the problem of spare time. Besides, spare time is usually pleasurable, not burdensome. Why do I work anyway, if not for the chance to relax and enjoy my free time?

*Answer:* I think that is what most of us would say. We would note that the so-called curse, free time is really a blessing, and, thus, we are doubly blessed; we have abundance and time to appreciate it. But in truth this extra time, time not biologically necessary to survive, is hardly the blessing we may think. It is a basic cause of the psychological ills that we all face. Without it, we are constantly just trying to survive. With it we are *re-creating* what was a real world for our ancestors and playing at survival—we work at our *recreation*.

# 3

# DISTRACTIONS AND DIVERSIONS

S PARE TIME (TIME FREED BY THE *ever-increasing efficiencies of civilization*) *can be filled in many different ways. Games of all types, alcohol and other drugs, and excessive physical and mental work are all important solutions to this problem.*

## Games And Other Pastimes

Civilization brought the secrets of abundance. Animals were herded and domesticated. Crops were seeded and harvested. Efficient energy attainment led to excessive energy and, hence, extra time. People have for the last 10,000 or so years been reaping this bountiful harvest.

However, people have not, in the last 10,000 years, or 500 generations, had a chance to genetically change in any dramatic way. We are still the wild hunting beasts we were before first successfully managing our resources. We still constantly search for and constantly need things to do. Thus, we play when not working; but both are necessary to sustain our physical and psychic needs.

Free time is time to play. If we define play as that which is not essential (to our survival), then it is more common the more efficient we become. Western societies have more free time than Eastern ones; an American worker devotes less effort securing his existence than an Asian peasant—but both have excess time to fill.

Games are legion. There are competitive individual and team sports. There are video and parlor games. Many of us enjoy gambling (at a casino or with friends). We partake of hobbies; we join clubs. There are as many activities as there are needs and desires. Frequently these pursuits mimic important aspects of work making us more efficient; but they all share the same essential purpose—they all help time to pass.

*Question:* Aren't you leaving out some major pastimes? Most of us watch TV, or go to movies, or attend events. We almost all read books, magazines, and newspapers. Many use the Internet for entertainment and knowledge (frequently texting information to friends and associates). These are all pastimes that we enjoy and that help us stay informed.

*Answer:* You are absolutely right. In a lot of ways, these are really productive and pleasurable uses of free time, and I am glad that you mention them. However, in general, these activities are not essential to our survival, so, in a sense, they can be considered games—constructive in important respects but, nonetheless, games.

As no one activity is truly more essential than another, after a while they begin to pale, and other pursuits, different things, are needed. As the mind is constantly looking, is continually active, and as there is nothing really important or life-sustaining to find, something that will ease this search is often sought.

## Alcohol

Societies long ago found that decaying plants contain an ingredient that changes the awareness of time, that decreases anxiety. Alcohol, the purified rot of fruits and grains, is this substance. In the civilized world, it is almost universally used. It matches a need, if not perfectly, at least closely enough. The requirement is to make unnecessary, anxiety-producing time pass; the solution is a drug that lets you enjoy doing little. It is a drug that allows silliness to reign.

*Question:* Are you stating that alcohol is a natural addiction, that it is as important as water or food? Are you implying that it is impossible to control a drinking habit?

*Answer:* No, I am not saying that. I am noting, however, that inasmuch as it fits a need, it will be used. Alcohol is a close match. It is inexpensive and easily made. In small quantities it is not dangerous and it does make time pass. Thus, alcohol is a much used drug, even a useful one.

## Other Drugs

There are many other drugs now used in this and other societies. Some use marijuana; it too, like alcohol, makes useless time pass. It similarly has been consumed for thousands of years and, in small quantities, probably is not harmful.

Other drugs are also widely utilized. Cocaine (with its derivative crack) seems to have been a recent rage. It gives a keener sense to time. It makes expectations appear to fit reality. One desires, and appears capable of doing, any number of what before seemed difficult tasks. The gulf narrows between desire and supposed ability. But, unhappily, this is an artificial narrowing, and once the high, the effect, wears off, cold reality returns. The sudden reappearance of this fissure, this abyss, is frightening, and the person often again uses the drug to make it disappear. Thus, cocaine can be an unsettling stimulant, dangerous and addictive to the user.

Caffeine, that common denominator of coffee, tea, and cola drinks, is also a stimulant. Again, like cocaine, it closes a gap between desire and ability. But unlike cocaine, it closes much less of a chasm. When it wears off, this breach reopens, but as it was not that huge, the side effects are marginal. Besides, caffeine is so widely consumed, perhaps more than alcohol, that it is tolerated and even condoned by our society.

Nicotine, also, is a very commonly employed drug. It is more habituating than caffeine, perhaps about as addictive as cocaine. It causes a diffuse, satisfying sensation, as if one had just finished an enjoyable meal. It acts as a vague tranquilizer; thus, it helps to lower anxiety. The only real problem is that the smoke it is inhaled with is dangerous, and we are becoming ever less tolerant of those who pollute others' air.

Opium and its many byproducts (heroin, morphine, and codeine, to name a few) are also regularly abused. They are very potent pain killers and anxiety suppressants. Endorphins (opium-like drugs) are also directly manufactured within the body. In cases of unrelenting pain, they are produced to allow one to move and survive. There are receptors in the brain that accept these endogenous opiates; thus, when a similar type of drug (morphine or heroin, for example) is taken, the body is already attuned to its actions. The response can be extremely satisfying, becoming an end in itself, easily leading to addiction. Thus, it not only helps time pass, it may become the full-time pastime.

Whether opium and its derivatives, in and of themselves, are harmful is another question. Probably when used judiciously and in small quantities, as when someone suffers from severe pain (a fracture, after surgery), they are safe. However, they do have significant addictive potential and, when abused, can be dangerous.

Many other drugs are extensively used in our society. But the basic fact of most is that they do something that is perceived as needed by their consumer. They tend to distort reality and ease the burden of empty time. Thus, like games, they have become a part of our civilized world.

**Question:** So you are saying that drugs (legal and illegal) are used to fill a need brought on by civilization. Therefore, as long as society endures, we will remain users.

**Answer:** I'm afraid that is so. They tend to be tools, adaptations of civilization. In nature, no animal can survive in a reality-distorted, drugged state. The birds that eat fermented berries fly and crash into branches. They are stunned and easy prey for predators. A drunken, doped person in combat with animals or other humans is a victim, not a survivor. Thus, drugs such as alcohol, opiates, or marijuana are dangerous in the wild state and have no long-term usefulness. It is only in a tame, domesticated society that persons can abuse these drugs.

*Question:* What other less addictive or at least less destructive means have we of filling this empty time?

## Work- and Thought- "Aholics"

*Answer:* Work is a great filler of time. Work, by definition, is productive; play is not. When enough has been done, allowing us to live as accustomed, we now have free time. If this becomes too troubling, some find work, above and beyond that needed just to survive, a refuge.

Those who utilize this option are our *workaholics.* They become addicted to work as an alcoholic is to drink or a substance abuser to drugs. The work, the earning of money, the accumulation of energy (power, wealth), is no longer required for real needs; it is done to fill empty, fretful time.

These persons become successful, occasionally fabulously so, as they can acquire many times what they use. If they were not addicted to constant work, if they did what was needed only to fulfill their needs, they would, like the rest of us, be faced with spare, unsettling time. As they find this too disturbing, they never finish; they keep acquiring. They act like George Eliot's fictional character Silas Marner, the miser, to whom money and not its use became most important.

Thus, to them, work becomes a full time process and the problem of spare time is shunted aside. These persons represent a positive force to humanity. They increase society's resources for their added accumulation, as it is not spent becomes part of everyone's prosperity. Henry Ford and Thomas Edison were such workaholics and their contributions form the basis for a significant part of today's world.

*Question:* You know, in a similar vein, those individuals who just spend unneeded time in contemplation (or writing) also fall into a like category. They could be called philosophers or dreamers, or, to coin a phrase—*thought-aholics.*

*Answer:* I think that you have aptly defined a group that seems to employ unwanted time in an appropriate manner. Obviously, you and I are members. Although we may consider our use valuable, we still are doing it to prevent the emptiness, the hollowness that superfluous time brings. By detailing our thoughts (for entertainment, knowledge, or any other purpose), we are, while attempting in a small way to enhance society, keeping our mental balance intact.

*Question:* So you are saying members of this last group also are comparable to other addicts, except that they, in the same way as workaholics, have found a method of filling time that may benefit society. Thus, those of us who play at games, who take drugs, or who work or ponder (and occasionally write) excessively are all after the same goal. Only for some it can be productive (increasing society's wealth and knowledge), for others it may be destructive (employing unnecessary drugs), but for many it is neither.

*Answer:* Yes, that is how I view people's actions; disparate though they may seem, they are all really done but for one reason. They are required to fill a gap that civilization has opened—a gap between energy attained and utilized. We, having become ever more efficient, are now dependent on diversions. That leads me to another very important method used to maintain our equilibrium—religion.

# 4

# RELIGION

R ELIGION IS ESSENTIAL IN ALLOWING us to cope with civilization and its inherent spare time; it helps to stabilize society. Monotheism is the contemplation of an underlying force—the cause and essence of all; this same force drives us to acquire, accumulate and advance civilization. Science attempts to understand how this force manifests itself—how it is perceived and works. In a sense, then, science is the cutting edge of religion.

## Monotheism

*Question:* You mention religion as a stabilizing force but appear to place it in the same category as drug and alcohol abuse. What are you attempting to do? Are you claiming that religion is but an addiction? You sound almost like a Marxist—religion, the opiate of the masses. Come now, religion is one of our greatest achievements; it is our attempt to understand the unknowable, to contemplate God.

*Answer:* Yes, I entirely agree. Religion is the pinnacle of our thoughts. It is the essence, at least in its pure form, of what we are capable of comprehending. If practiced in a tolerant way, it is the best guide we have to a proper society. Yet religion, like the previous diversions, is, in part, also a consequence of civilization. It too, like excessive work, drugs, or games, becomes a way to dampen the anxiety of free time.

Organized religion, as presently followed, most likely did not predate civilization. What did prevail was not as observed today and almost certainly would not have been monotheistic. Anyway, we need not speculate, for we have no way of knowing what prehistoric humans practiced; once we became civilized, religion as we understand it came into existence.

Early religions worshiped physical things; these totems were given supernatural powers to help or hurt other individuals or opposing tribes. As societies evolved, so too did these ritual objects. When a rock no longer held mystery, a mountain became a deity. In time, even the mountain became comprehensible, and the heavens then encased the Gods. Finally, the ultimate concept, monotheism, was conceived. God, the unknown, the cause and essence of all things, became our focus.

Monotheism, the worship of one creator, of one spirit, when established became an essential aspect of all psyches. It is, in mental terms, precisely what the world is in physical terms. It is an exact depiction of the force that maintains us, that keeps us growing and evolving.

**Question:** You spoke before of our innate need to amass and acquire—our hunger for more. Now you describe it as a force that preserves us, permitting us to grow and advance. What are you actually trying to explain?

**Answer:** There is a force, manifested as gravity on a global scale, which presents as a continuous drive of accumulation. It is perceived as a constant competition for energy. It holds sway over all things. It allows for existence. As already noted, we will make an attempt, soon, to try and understand it, but let us continue with our current train of thought.

## Science—Religion's Handmaiden

As societies change and grow, even monotheism, the glorification of God, must evolve. Minds keep seeking, and different avenues appear. People gradually master science (the logical accumulation of knowledge), and newer and ever more effective means of acquisition are grasped. However, human thought always returns to the same basic concept, the ultimate source; science, thus, as the cutting edge of today's religion, becomes but a more rational and efficient search for this essence.

**Question:** Wait a minute! I always thought science and religion were in conflict, that they were not compatible.

**Answer:** Many of us, like yourself, think of science and religion as opposites: One concerns what is real, the other what is imagined. However, they both have the same purpose. Science is but the logical extension of religion. It shows how, but its goal is what is. It is attempting to clarify monotheistic concepts, and today's science will be incorporated into future religious teachings.

**Question:** The idea of science and religion as really the same still seems strange. However, when you see cosmologists trying to explain the origin of the universe, you begin to understand how they and our religious leaders may have the same inspiration and vision. I guess that is what you are alluding to?

**Answer:** Yes, that is certainly a part of it; religion, as the worship of a single God (monotheism) is the basing of all on a specific source (a force that controls and permeates all things). It is very similar to our scientists trying to unify all forces (gravity, electromagnetism, and the strong and weak interactions). It was really the drive behind Einstein's idea of the universe; for Einstein, although not openly orthodox in his practice of religion, was deeply spiritual.

The theorists who now try to understand these problems are just following in his and many other great scientists' paths. As more is known, as more of this fundamental force is uncovered, science and religion will grow closer. Religion as it evolves will assimilate these findings and change.

This process has already occurred in Jewish and Christian thought with philosophers like Maimonides and St. Thomas Aquinas; both endeavored to incorporate *scientific* (Greek Aristotelian) concepts into religion. At first their ideas were shunned, but, in time, they became the mainstream of religious belief. Religion—monotheism—will always rule supreme as the ultimate answer is unknowable. The final quest will always be couched in religious terms, but access to it will, for the foreseeable future, be viewed scientifically.

## Maximizing Energy

*Question:* You seem to have quieted my concerns about science verses religion. Why don't you get back to your original thought about why religion is of such fundamental importance to people.

*Answer:* Thanks for keeping me on track. I was trying to show why religion has such a strong hold on civilized beings. Again, I go back to my original point. Civilization leads to empty time, time not needed for survival. This time hangs heavily; it must be filled. When all our myriad pursuits fail, when the mind, active as always, continues to seek, then religion allows for a comprehensive resolution. Religion, by calling on the *infinite,* fills this immense void into which one can fall.

There cannot be anything greater, as it is boundless and incorporates everything. It, therefore, can be a final stop for a searching mind. We are, as are all things, designed to maximize energy. God now becomes the ultimate energy or force. Thus, we are designed to believe in God, and God becomes a safe haven for an anxious, searching mind.

Monotheistic thought, the concept of God, is built into the design of all minds. *Organized* religion is born from the awareness of this fundamental need. It allows for the control of energy (food, money, power) by a group (priests) who can mold others by channeling their anxieties and fears. Thus, a society with an organized monotheistic religion becomes stronger and more stable as its most fundamental needs are addressed. Religion, once organized, can seem appropriate or not depending on one's personal view; but the reason for its existence is basic to the design of all things.

**Question:** So are you saying that leaders of organized religion are taking advantage of a need that is inherent to all and using it, just as dope peddlers use narcotics, to control others? Therefore, are you not tarring religious leaders with the same brush as drug dealers?

**Answer:** Well, not really. War in the name of religion has caused great personal suffering. Religion has certainly caused a lot of hatred; just look at the Jew and the Muslim in the Middle East, or the Catholic and the Protestant in Ireland; however, religion, as it is not a true physical addiction but really an intellectual pursuit, can also bring out the highest strivings and thoughts of a people. Many of our greatest philosophers were religious leaders.

Whether or not one agrees with any established religion is one's own prerogative; that religion exists is fundamental—hopefully it will generate more good than bad. Anyway, I am certainly not a Marxist, and I do not feel that religion is similar to opium, for it can expand and enrich our minds creating more wondrous thoughts for all.

**Question:** I really did not think you were so against religion, organized or not. My earlier allusion to Marx was partly in jest. However, did he not try to foster his economic solution—communism—as almost a sacred mission? Do you see any link between Marx's interpretation of economics, with his concept of the inevitability of communism, and a religious quest?

# 5

# POLITICS

C*OMMUNISM, ESSENTIALLY A RELIGION without God, is a failed political system. Capitalism is successful, as it unleashes the underlying force that gives rise to acquisition and accumulation. Fundamentalist societies, those without democracy or capitalism, are inherently unstable and cannot succeed in competition with more vibrant and productive ones.*

*The most effective societies, then, are democracies with a strong capitalist and spiritual foundation, such as the United States. Modern post-war Europe faces two major problems—loss of religious fervor, and suppression of industrial vigor. The only true rivals to the United States, therefore, are other similar competitive societies with strong underpinnings of faith.*

## Marx's Mistakes

**Answer:** To properly answer your question about Marx, I must, first of all, state that I fundamentally disagree with his interpretation of history and economics, specifically his notion of the inevitability of communism. In fact, I believe in just the opposite, the inevitability of capitalism. Capitalism, as we understand it, is a system of competition, with the winners being the most efficient organizers and distributors of products. In the long run, societies that allow this practice to thrive will outproduce and overwhelm those that impede it; just consider the failed Soviet empire.

What Marx should really have stated was simply that a population will almost always fall into three fundamental groups—the *have enough*, the *have some,* and the *have none.* Those with *enough* control most of the means of power: businesses, land, wealth. They wish to *conserve* it. They are the Republicans, the Tories, or the Right Wing.

Those with *some* would like to obtain additional riches but, by already owning a portion, have something to lose in a nasty fight. They do not wish to wage an actual war—they simply want more. They generally are workers with relatively stable jobs who, having tasted some wealth, desire more but not enough to risk all; they wish to *liberate* the resources of others'. They are the Democrats, the Labor Party, or the Left-Wing.

Finally, the group with *nothing* has nothing to lose and is willing to fight for what it needs and wants. This group is comprised of the poor with no hope—slaves, starving workers, landless farmers. In a stable society, this class is small, for stability means that most have some modicum of wealth, no matter how little.

## Communism And Chaos

In an unstable society, however, such as Russia during World War I with starving peasants and displaced workers dying in a war that was to them of no fundamental benefit, or in any small *banana republic* such as those in Central America where the poor are truly destitute and without hope, this last class can grow. This group we call Communist; however, Christians in ancient Rome made up the same class. If a society is reasonably run, this party remains small and ineffectual, but if a society is unsettled it may expand and conquer.

*Question:* Well, using your own criteria, don't we see a stirring today in the Occupy Wall Street crowd? Couldn't this be the beginning of a more disruptive movement?

*Answer:* I do not think so. These are not starving or desperate people. Their current lament is really but the incoherent cry of the liberal wing of the Democratic Party and, most probably, will be so assimilated. In a like manner, at the other extreme, today's Tea Party is merely stating the fiscally conservative ideas of the Republican Party and should ultimately be retained there.

To get back to Marx, he was mistaken when he noted that communism would take hold in industrialized Europe. It could only start among a wartorn, starving people, as in Russia or China. However, once it seized power, it behaved as did any other ruling group. If we then take Marx's other precepts, we have very much the same teachings as seen in fundamental Christianity—the good of all, the sharing of wealth—without the cohesiveness brought by a belief in God. In a battle against those with religious faith a movement without its essence, Godless Marxism, will certainly lose to God-fearing forces.

In a confrontation, society to society, the one based on competition and capitalism will, in the long run, overwhelm the one that is not. The caveat is, of course, the *long run*, for in a short period other factors (territorial advantage, strength of forces) all need to be considered.

## Religious Fundamentalism

*Question:* I see what you think about the failure of communism. When it came up against a free, competitive state with a faith in God at its core, it was defeated. However, today we see another potential threat—a resurgence of fundamentalist Muslim belief. What do you see happening in this new arena?

*Answer:* Again, the same arguments apply. The United States is a free and competitive society with a deeply ingrained belief in God. Freedom with capitalism allows for a maximizing of goods and services, including armaments. Thus, we are the best armed and defended society in the world. Our fundamental trust in God is the cohesive force that makes us so powerful.

The Islamic world is really split into two major groups. There are the Asian countries, relatively free and democratic, with strong religious foundations, and as long as they remain so (allowing their citizens to fully compete, to acquire and accomplish what each desires), they will be our equal and can succeed as have we.

The other segment of the Islamic world, the nations of North Africa and the Middle East, although having equally strong religious underpinnings, are generally not free and democratic. Their citizens often are not allowed to pursue their basic desires. Their societies are not nearly as competitive; frequently their women are excluded from normal commerce. Their fundamental belief in God is a potent and cohesive force; however, they are competing against us with one hand tied and, if they do not change (become democratic) will fail.

Perhaps we are beginning to see a stirring with the partial democratization of Iraq and the turmoil spreading throughout the area—the *Arab Spring*. This movement, if it continues and envelops the entire region, should strengthen all these states and make them able competitors on the world stage.

## Europe's Dilemma

**Question:** But today's battle line is not just seen in the United States; it is also seen in modern Europe. Can Westernized European civilization survive?

**Answer:** Today's Europe is wholly democratic; two world wars brought this about. However, much of Europe is economically noncompetitive. A lot of production is State-owned or controlled and, thus, inefficient. Goods and services cannot be as readily produced as in the United States; armaments are not as plentiful.

Following the two great wars, the people opted for security and statism—government and labor union control of production. The people also seemed to lose some of their religious fervor. We see this trend both in the cathedrals of Western Europe, which often seem little more than museums, and in the constitution of the European Union, which specifically omits any reference to the role of Christianity in the formation of modern Europe.

For many years, Germany was Europe's problem; now, Europe has become Germany's. By losing the war instead of enslaving Europe, Germany now has to support it. If Germany's interest wanes, the vaunted European Union will totter. The countries of Europe's southern tier are essentially noncompetitive and potentially bankrupt. Without that region becoming more industrious or religious, the Union can fail, and we may find, in the foreseeable future, a resurgence of Islamic faith and a schism between a Westernized Germanic enclave and a Muslim Mediterranean one. Thus, today's Europe, although appearing on the surface as a democratic, capitalistic, God-fearing society, has a much softer underbelly.

## Asian Competition

**Question:** But what about competition from Asian societies, specifically China, India, and Japan; can't they potentially overwhelm us economically and even militarily?

**Answer:** You have lumped three distinct entities together. China was an autocratic, previously static society that is slowly loosening its top-heavy controls. As it becomes freer, and its people are allowed fuller rein, it will continue to expand and prosper. It is a very large country, filled with intelligent people, and it could eventually become the preeminent force in the world. Its spiritual and religious foundations go back millennia, and its present government appears to be more willing to encourage these fundamental and important beliefs.

India and Japan are freer societies, places where individuals are able to compete for status and wealth. Both have religious underpinnings, even if not necessarily monotheistic, and both, in their own way, are true competitors of the United States.

The United States needs to maintain its essence—democratic, capitalistic, and God-fearing—to help its traditional partners and to prevail. Do not forget our motto is *in God we trust*. As long as our country maintains its fundamental base, other societies, those either God-abandoned, static and essentially socialistic, or God-fearing but still stagnant and authoritarian, cannot triumph. Thank God for America.

**Question:** Well, that certainly tells me what you think about politics and religion. However, you have been constantly alluding to a force that seems to permeate all things, that appears to govern our actions and causes things to exist. Perhaps now you can attempt to explain what you mean.

SECTION II

# Einstein, Hubble, and the Big Bang

INSTEIN'S THEORY OF RELATIVITY is discussed from the viewpoint of the constant, unreachable speed of light. Since all velocities when compared to light are the same, distortions in length and time as objects speed up can be more easily visualized. Analogies are given using the concepts of 1D and 2D universes. Our 3D universe is, then, shown to be but the surface of a 4D sphere, and distortions in length as we gaze out toward the periphery are explained.

Hubble's concept, an expanding universe, is discussed based on his finding of an increasing redshift with distance. This led to the Big Bang theory (BBT). Pros and cons are given; mainly, cosmic microwave background radiation (CMBR) versus an inability to understand 95 percent of what exists (dark matter and dark energy).

Dark energy (the hypothetical cause of an even greater expansion of the universe) is assumed to exist as entities were discovered farther away than described by Hubble's theory. However, when using a 4D explanation (simple tangents), the real distances are found and they more closely correlate with current observations. Therefore, dark energy (the unknown expanding force of the universe) is found to be but an illusion, and a new paradigm, based on a higher dimension, is proposed.

CMBR, the most important reason for the acceptance of the BBT, is shown to be the enormous stretch at the very edge of the universe. Thus, the Noble Prizes awarded to both the discoverers of CMBR and dark energy were well deserved but given for the wrong reasons; they should have been for refuting, not proving, the BBT.

# 6

# EINSTEIN'S WORLD

THE MICHELSON–MORLEY EXPERIMENT led to the strange finding of the constancy of the speed of light. Einstein used this result as an axiom in his special theory of relativity. Because of this, he established that distance contracts and time expands as one travels ever faster—at speeds approaching that of light.

To more easily understand this, one needs to visualize all velocities from the viewpoint of light. To light, all speeds are the same, as, no matter how fast an object travels, light always moves by it at the same constant rate. Using this concept, we can show why the alterations found by Einstein actually occur—and these distortions, these changes, are the basis of how the universe really works.

## The Universe Prior To Einstein

*Answer:* What you are asking is difficult to explain. It is what Einstein sought as the single unifying force. It is almost like describing God. Let me try to place into perspective the world as it appeared to Einstein and perhaps, then, we can start to understand.

Around the end of the nineteenth century, physicists were trying to interpret the results of experiments performed by two scientists—Albert Michelson, and Edward Morley. They had been trying to better characterize *ether*, a theoretical substance (the medium or essence through which light was thought to travel) that was considered to permeate all things. In their investigations, they tested the speed of light shining both with and against the direction of the Earth's motion. In both cases, the speed was exactly the same.

This confounded classical physicists. Until this time, scientists had believed that ether was a material (such as water or air) that could transmit waves. However, if an actual substance were encountered by a planet as it moved in orbit, and if a wave in that substance were to be intercepted, then the speed of that wave at impact would be greater if hit head-on than if hit from behind.

*Question:* Oh, I see. You mean like a head-on versus a rear-end collision.

*Answer:* Exactly: There is always much greater force when two objects meet head-on. This force is due to the combined speed of the two entities. In the Michelson-Morley experiments, the speed of light should have been greater when opposing the Earth's motion, but it was not; the speed was the same in both directions.

Prior to 1900, physics had largely been based on Newton's ideas. He had described a force, gravity, that acted upon every particle in the universe. If one knew the mass of an object, one then could measure its gravitational attraction. Newtonian thought was the recognized way of viewing the world. It was accepted because it was almost always correct. Occasionally an inconsistency was noted, but in the great majority of cases Newtonian concepts explained the facts.

*Question:* But what does this have to do with the experiments we were talking about before—you know, the ones showing how strange the speed of light was?

*Answer:* I just put that in to help give a little perspective. Newton was not the only scientific voice prior to 1900. Michael Faraday, in the early 1800s, was the first to describe electricity and magnetism as forces permeating space. However, he was not a mathematician and could not give formal proofs of his ideas.

About 1860, James Clerk Maxwell, a Scottish physicist, was able to rigorously prove that electricity and magnetism traveled together as a wave, just as had been shown for light. This was a major scientific advance; now there were two fundamental aspects of nature, gravity and electromagnetism. Physicists used these basic forces to explain the facts as they became apparent.

## Relative Speed Verses Light Speed

Thus, when Einstein first began to publish his work, physics was founded on the two rocks of Newton and Maxwell, or gravitational and electromagnetic force. Most things were understandable using these two concepts, but not *all*. The great unexplainable was that nagging fact that light (or electromagnetic energy) does not act like anything else.

All other things travel at speeds relative to their surroundings. Thus, if an individual walks at 4 mph and a bicyclist pedals by at 10 mph, the cyclist is going faster than the pedestrian by 6 mph (10 minus 4). If a car speeds by at 60 mph, the car moves past that person at 56 mph (60 minus 4). If a jet plane soars past at 600 mph, the jet is traveling at 596 mph with regard to the walker (600 minus 4). All physical things, thus, travel at speeds dependent on others. Light does not.

Therefore, if one is walking or running or traveling in any conveyance (at any velocity whatsoever), and if light passes that person in either direction, light will be going at the same rate as if one were standing still (even if one were moving at 99 percent or more of the speed of light). This did not make sense to classical physicists, and no matter how they struggled, they could not incorporate it into their theories.

**Question:** I feel for these scientists, for it really does not make any sense to me either.

**Answer:** It does not make sense to us just as the Earth being round appeared nonsensical to our ancestors. When told the Earth was round, a common reply was, If that were so, why then does one not fall off when standing on the opposite or bottom side? Similarly, the idea of a moving Earth and a stationary Sun was illogical to most. When told that the Earth rotated, the usual retort was, If that were true, why then does one not feel this motion?

Thus, things that we now undeniably know are correct, and accept as fact, at one time in the past made no sense either. However, they do now; they have been incorporated into scientific thought and distilled into everyday belief, and have become *common sense*.

*Question:* So what did Einstein do? What was he able to make of those strange, speed-of-light experiments?

## Einstein's New Postulates

*Answer:* He was able to explain them by setting up two new postulates. *Postulates* or *axioms* are accepted truths; for example, an axiom in geometry is that the shortest distance between two points is a straight line. What Einstein proposed was that the speed of light was constant no matter at what speed the viewer, or the source of the light traveled.

He simply took the results of the Michelson-Morley experiments, and other similar contemporary ideas, and, without explaining them, incorporated them into a basic belief in his new theory. He also assumed that the laws of physics were the same in all locations and for all entities (if there were no accelerating forces present).

*Question:* So you are telling me he did not explain those experiments, he just used their conclusions as a basic new truth?

*Answer:* That is correct. But once that idea became fundamental, a whole new approach was possible. This led Einstein to his initial, or special, theory of relativity. It grew out of the two postulates or assumptions that he put forward.

## Distance Contraction

To appreciate Einstein's special theory of relativity, we must go back to his proposal that the speed of light is constant no matter at what speed you travel. Thus, if you travel at 1 mph or 1,000,000 mph, light appears to be traveling past you at almost 700,000,000 mph (186,000 mps or 670,000,000 mph). Even if you were to travel at very, very rapid speeds—500 or 600 million mph—speeds which are utterly unobtainable with any known technology, light would still be passing you at nearly 700,000,000 mph.

Thus, all the velocities at which you move appear to light as the same; 1 mph and 1,000,000 mph become equal. Now, we know that, in the real world, 1 mph is significantly different from 1,000,000 mph. But if we *assume* them to be equivalent, as they are with respect to light, then we must suppose that something fundamental happens to the way we measure miles (or any distance) and hours (or any time) the faster we travel. Therefore, if 1 mph is equal to 1,000,000 mph, we could write an equation as follows:

$$SLOW = FAST$$
$$1 \text{ mile/hour} = 1,000,000 \text{ miles/hour.}$$

Then we can say that:

$$1 \text{ } slow \text{ mile} = 1,000,000 \text{ } fast \text{ miles}$$

as they are both distances traveled in the same period of time—one hour.

Now, a slow mile can only equal 1,000,000 fast miles if the fast miles are much smaller than the slow miles. Thus, the fast miles, in this example, must be 1/1,000,000th as large as the slow miles. Or we can say, if we accept Einstein's postulates, that distances get smaller or contract as we speed up.

**Question:** But how could this be true? I have traveled in jet planes at 600 mph and have not become 1/600th as large as before I left.

**Answer:** You are right. We are not saying that, at normal speeds, compared to light speed, any noticeable change would occur. The change at relatively low velocity is so minute that we would be hard-pressed to note it even by very precise measurements. But as the velocity increases and grows closer and closer to that of light, the small, subtle changes become more obvious, until major distortions occur. I only used 1,000,000 mph to denote a very fast rate. In reality, it must be much, much faster still to show significant changes.

*Question:* I seem to get some glimmer of what you are saying. It seems that all velocities compared to light are really the same, and that, when we go faster, the distances we travel shrink in length. When we travel through space at fast rates we travel *fast–small* miles and at slow rates *slow–big* miles.

*Answer:* Well, it is something like that, but there is more weirdness yet if you agree with Einstein's postulates. Just as distance changes time also changes as we speed up.

## Time Expansion

*Question:* I think I have heard of this too, although I am almost afraid to say. I have read science-fiction tales of people traveling to distant stars at very rapid speeds who, upon returning to Earth, had aged only several years, whereas everyone they had known on Earth had passed away years before; but I always thought these stories were just that— science fiction. I guess they are real, too?

*Answer:* They are based on facts, although whether we could travel fast enough to make them occur is another thing. But just as we saw distance altered, we can see time change. It is really simple to understand if we assume the same postulates to be true.

Again we are using Einstein's postulate that the speed of light is constant. Thus, no matter at what speed you are traveling, light is traveling at the speed of light as compared to you. Therefore, (to light) all velocities are the same; hence, (to light) 1 mph equals 1,000,000 mph, or as we already stated:

$$SLOW = FAST$$
$$1 \text{ mile/hour} = 1,000,000 \text{ miles/hour}.$$

Before, when we assumed the hours to be constant, we saw that the miles contracted (as we increased speed). If now we consider the miles to be constant—that is, they do not contract at a high velocity—then the hours must change. If we divide both sides by 1,000,000 we have:

$$\frac{1}{1,000,000} \frac{\text{mile}}{\text{hours}} = \frac{1,000,000}{1,000,000} \frac{\text{miles}}{\text{hours}}$$

or

$$\frac{1}{1,000,000} \frac{\text{mile}}{\text{hours}} = 1 \frac{\text{mile}}{\text{hour}}.$$

Now, if miles are the same at slow and fast velocities, we then have:

$$\frac{1}{1,000,000 \text{ } slow \text{ hours}} = \frac{1}{1 \text{ } fast \text{ hour}}$$

or, if we cross multiply

$$1,000,000 \text{ } slow \text{ hours} = 1 \text{ } fast \text{ hour}.$$

What we are saying is that, in one million slow hours, we have traveled the same distance we would have traveled in one fast hour (as we are keeping the distances the same). Therefore, those of us who are traveling at a fast rate (and this velocity needs to be almost 700,000,000 mph) will have aged only one hour, whereas, those traveling slowly will have aged one million hours (or 114 years). Thus, time slows as we speed up; those of us who travel faster will age at a decreased rate. Again, these changes only begin to be appreciated at very, very rapid speeds, at those approaching light.

## Space–Time Distortion

*Question:* I don't know. I don't like math, even simple math, and what you did looks like a math trick to me.

*Answer:* It is not entirely a trick. I was trying to show that all velocities (ratios of distance/time) were the same with regard to each other, as they were always the same concerning light.

*Question:* If this is so, why don't we sense it?

*Answer:* We are almost never aware of it because major distortions only begin to occur at extremely rapid velocities, those that approach the speed of light. Almost nothing that we experience, or can travel in, moves at these speeds.

*Question:* If we only rarely measure it, and if it seldom is important in our lives, why make it such a big deal?

*Answer:* The reason is that, although we do not routinely sense these changes, the so-called distortions or curvatures in space and time, they are fundamental to our universe. They are the basis of why things occur. Let me give you examples using one- and two-dimensional analogies: the worlds of Lineland and Planeland.

# 7

# OTHER DIMENSIONS

USING A MODEL WITH ONE- *and two-dimensional "make-believe" universes allows us to visualize how a higher dimension would be perceived—essentially as a continuously increased stretching of space the farther out one looks. Our three-dimensional world is then "bent" toward a fourth dimension as we try to understand this concept (a stretch or expansion in space).*

*Our fabricated two-dimensional world (Planeland) is also used to show how higher three-dimensional spheres would be seen—as circles within circles spreading to the borders of that universe. In a similar manner, our real three-dimensional world visualizes non-describable, higher-dimensional "orbs" as ever expanding concentric spheres to the limits of understandable space. As every fourth-dimensional object extends out to the very edges of our world, each must include all others— they are all intertwined, they are all entangled.*

## Lineland

Lineland is a one-dimensional universe. It exists solely as a straight line, and its inhabitants are simply segments of that line. All they know or can imagine is there, on a line. Everything that exists must be incorporated into that narrow path, as it is their entire world. If that supposed *straight* line were really *curved* (a two-dimensional circle) how would its inhabitants view that higher plane?

*Question:* I don't know; why don't you just show me.

*Answer:* Linelanders would consider that second-dimensional distance a stretching or elongation of distance in their own dimension. They could only conceive of the curved line as straight; and distance, therefore, would keep enlarging the farther out they looked. If we were to draw a quarter circle to represent that second dimension, the equal lengths (a–e) would appear to keep increasing (1–5) until they reached infinity or disappeared. Let me show you.

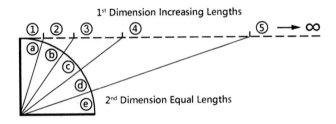

Thus, the *farther* one gazes into a higher dimension, the *greater* the length sensed in the lower one.

*Question:* Are you saying that, from any point in the first dimension, distances in the second appear expanded, and the farther out, the greater that change?

*Answer:* That is exactly what I am saying. What is really the *same* size keeps lengthening as one peers ever farther into the higher dimension.

*Question:* So you are using the simple example of a one-dimensional universe in two-dimensional space to show that the higher dimension is sensed as changes (elongations) in distance measurements.

## Planeland

*Answer:* That is correct. Let us now go to a *two*-dimensional world—Planeland. Just as in Lineland, all Planelanders know, or can conceive of, is their own universe—a flat two-dimensional surface world. But if Planeland were the surface of a *three*-dimensional sphere how would they perceive that object?

*Question:* I guess there would be a lot of distortions as the inhabitants tried to visualize what existed in another dimension.

*Answer:* You are right. From our perspective, as three-dimensional beings, we see spheres and flat planes; but the Planelanders, who cannot visualize or understand this three-dimensional universe, see only concentric circles. Let me draw it.

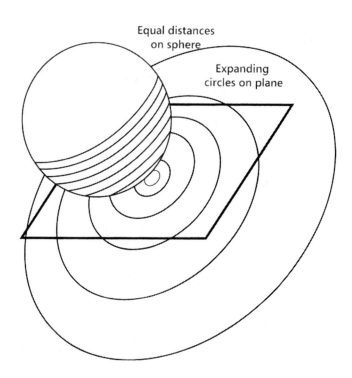

Equal distances on sphere

Expanding circles on plane

If we draw any sphere with equidistant surface markers (inches, meters, or whatever), these lines going about the sphere would be seen by Planelanders as circles separated by ever-increasing spaces starting at a central point. Just as Linelanders (upon gazing into the distance) saw increasing space between markers, Planelanders see similar changes as ever-enlarging concentric circles. These circles of increasing diameter go on and on until finally they disappear completely into infinity.

*Question:* So you have used the analogies of one- and two-dimensional universes (Lineland and Planeland) to show how length increases the farther out one peers. Where does that leave us, the residents of three-dimensional space?

## Three Dimensions—Our Land

*Answer:* I am glad that, so far, you grasp my ideas. I will try to show you how we can visualize a higher, or fourth dimension, and how we are affected by its existence. Let me once again use drawings to help us understand my concepts.

In our drawings, we take a cube of space (three-dimensional) and

Cube

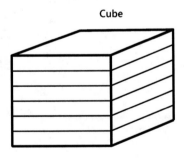

put equidistant lines (as distance markers) onto its surface. We then flatten it front to back, so that it forms a tablet. It is still three-dimesional,

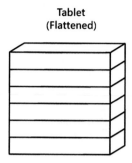

Tablet
(Flattened)

only foreshortened. Finally, we bend it in a fourth, or other direction.

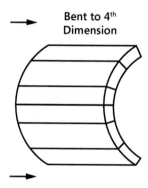

Bent to 4th
Dimension

Obviously, as three-dimensional beings, we cannot visualize this higher dimension. However, we can begin to see distortions in our equidistant markers due to this bend and, as we go farther away from the center of the cube's surface, toward the periphery, the distances between the lines widen.

**Question:** I see what you are trying to show. From any point in our three-dimensional world, the farther out we look, the greater the same distance would appear. These changes in length are simply due to the geometry of trying to visualize a higher dimension in a lower one. But in what direction have you bent the tablet? There is only up/down, right/left and front/back. Where is this other, or *fourth*, direction?

## Visualizing The Fourth Dimension

*Answer:* We really cannot comprehend this higher dimension, as we are inhabitants of a universe of just three spatial dimensions. All we can say is that, if we were bent toward that direction, distortions in length would be seen. However, if we continue the drawings, we may be able to visualize what is inherently unknowable. Thus, we now take this bent tablet and place it on the surface of a sphere.

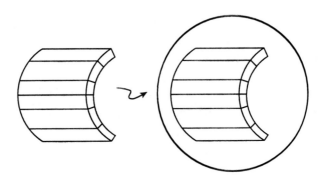

The sphere can be said to be fourth-dimensional, as its surface is composed of bent and distorted three-dimensional cubes. However, my drawing is simply a representation (using a three-dimensional sphere); we just have to try and imagine that it has four dimensions.

Now, let us say that our fourth-dimensional sphere is surrounded by other, similar fourth-dimensional spheres (or hyperspheres). How would we as three-dimensional surface inhabitants sense these other entities?

*Question:* This is getting too complicated for me. The Lineland and Planeland analogies were relatively easy to visualize. Size changes and circles within circles made sense. However, now you are bending my mind a little too much.

*Answer:* I admit it is hard to visualize, but let us just use our prior concepts and extend them a little. Remember, in Planeland a three-dimensional sphere from any point was visualized as ever-enlarging concentric circles out to infinity. So to transfer these concepts to our world, a world of three spatial dimensions, four-dimensional hyperspheres would appear to us from any point as ever-enlarging three-dimensional concentric globes out to infinity.

*Question:* I begin to see what you are saying. When inhabitants of a lower dimension view a higher one, they do not see that dimension, they simply see continually lengthening intervals; and the farther they look, the greater that stretching grows, until it finally disappears into infinity. Thus, since we are three-dimensional, we visualize hyperspheres as orbs within orbs. But how can such things actually exist? Don't forget we are made of *matter*. How can we have multiple, ever-enlarging concentric spheres if they are all solid? It makes no sense.

*Answer:* The answer lies in Einstein's famous equation $E = mc^2$ (energy equals mass [or matter] multiplied by the speed of light squared). Since the speed of light ($c$) is a constant (300,000 km/sec), energy and mass are equivalent. Thus, tangible (matter) spheres are not required. We just need ones made of energy, and, as in the analogy of Planeland, these spheres continue out to the edges of the universe—to infinity.

Also, every set (in our world) of concentric spheres (from each individual hypersphere) would, by extending to infinity, encompass all others; they would all be intertwined—they would all be *entangled*. This concept will be important later when discussing *quantum weirdness*.

*Question:* Okay, I understand what you are trying to show; however, you will have to clarify it a lot more before I am ready to accept it.

# 8

# REDSHIFT AND THE BIG BANG

T HE INCREASING REDSHIFT SEEN AS *one gazes ever farther into space was felt by Hubble to be a Doppler effect. This implied an increasing velocity the farther out one looked. The Big Bang theory, was then proposed to explain these results; in essence, the universe began as one inconceivably dense point, which exploded outward and led to all that we now see or understand.*

*The discovery of cosmic microwave background radiation cemented this theory, and the additional findings of the hydrogen-to-helium ratio, and the immaturity of distant, primitive galaxies, convinced most scientists of its validity. However, very early, extremely rapid inflation and dark energy are both needed to make it work. So today the real problem with this theory is that we understand only 5 percent of the entire universe—95 percent is pure conjecture.*

## Redshift

**Answer:** Let us discuss other theories, and you may then see what I am getting at. You have, of course, heard of the Big Bang?

**Question:** I most certainly have; everybody has heard of it. The Big Bang theory states that the universe started a long time ago as a single, very dense point and exploded outwardly to what it is now.

*Answer:* You have the gist of it, but let me try to explain some things to give you a fuller picture. The Big Bang theory is based upon astronomers' findings, dating back to the late 1800s, of anomalies (or redshifts) frequently seen in the spectral lines of the light coming from objects in deep space. In the 1920s, Edwin Hubble, probably the most proficient astronomer of his time, with the aid of what was then the world's largest telescope (the 100-inch instrument at Mt. Wilson), was able to record these changes out to distances of 50 to 100 million light-years. He found that, the farther into the void he looked, the greater the redshifts appeared.

*Question:* What do you really mean by a redshift?

*Answer:* Light is composed of many different wavelengths or colors. Plain white light really consists of red, orange, yellow, green, blue, indigo and violet hues. An easy way to remember this is the mnemonic *ROY G BIV.*

Red has the longest wavelength—around 750 nanometers (750 x $10^{-9}$ m), and violet the shortest—about 400 nm. These are the colors that we can see. However, there are innumerable wavelengths that are not visible; those longer than red make up the infrared spectrum, and those shorter than violet the ultraviolet region.

Now, light, because it is a wave, shortens or lengthens as its source approaches or leaves; this is called the *Doppler effect.* A good example is the siren of an ambulance whose pitch (or frequency) is higher while it travels toward us and lower as it departs. The pitch that we hear is simply a measure of the length of the sound wave. The higher it is, the more waves pass by us each second, thus the shorter each wave; the lower, the fewer waves per second, thus the longer is each wave.

*Question:* I see what you are saying. A redshift means that an object is increasing its wavelength and moving away from us; therefore, the lowering of the siren's pitch means that the ambulance is leaving.

## Expansion Of Space

*Answer:* You are correct; however, also remember the *longer* the stretch in the wavelength (the lower the pitch or frequency), the *faster* that entity is leaving. What Hubble saw as he looked farther and farther out was that most objects appeared redshifted, and that the greater the observed distance, the larger was that shift. He deduced, therefore, that all objects were moving away from our point in space, and at increasing velocities—the farther out the faster they receded.

Now, Hubble could only discern astronomical masses out to around 50 to 100 million light-years. Today it is thought that the universe extends outwardly about 13.80 billion light-years (current best estimate— 13.798 +/- 0.037 billion). Thus, Hubble could only see less than 1% of the entire cosmos. However, with his limited information, he concluded that it appeared to be expanding, at ever higher velocities, the greater the distance from any spot.

Some cosmologists then theorized that, if the universe was expanding in all directions, it had to have started at one point. A primal explosion was postulated, which later was derided by opponents as a "Big Bang." The name stuck and became popular; and the theory accepted by most today is that the universe started from an infinitesimal location (in space and time) and has expanded outwardly since.

*Question:* You mentioned that the name—Big Bang—was originally given by the theory's detractors; if they disagreed, what were their counter-arguments?

*Answer:* In the 1950s other scientists, using the same data first supplied by Hubble, came to the conclusion that, although expansion was real, the universe did not start from a single point. They called their concept the "Steady State" theory and maintained that the universe had always existed in its current form but was constantly expanding as new material (hydrogen atoms) continually formed. These conflicting theories were hotly debated; the less concrete the information, the more vigorous the controversy.

## Cosmic Microwave Background Radiation

However, in 1964, two scientists (Arno Penzias and Robert Wilson) working with a radio telescope at Bell Labs found a uniform background *noise* present throughout the universe. Their finding occurred truly by chance as they were merely trying to cancel what they thought was *static* present in their instrument. As they were unable, using all their available resources, to rid it of this noise, they reported their finding as real.

Big Bang supporters had theorized, some time before, that, as the universe cooled from its initial *explosion,* existing matter would congeal from a plasma state (with chaotic disconnected protons and electrons) to the way it currently appears (atoms and molecules). They felt that, as it cooled, the *fog* of plasma (through which the initial *light* of the primeval explosion was barely visible) would lift, and the event would then be seen. Over eons of time, this light would have expanded or shifted its wavelength more than 1000 times, into and beyond the red spectrum (going from visible light in the hundreds of nanometers to invisible microwave radiation measured in millimeters and centimeters).

Since the microwave background radiation so closely fit the theory of the Big Bang, and as it could not be explained by Steady State adherents, most began to accept the Big Bang as correct.

## Genesis

**Question:** You know the Big Bang almost sounds like the Biblical story of Genesis, with God creating the universe from nothing. Is there any connection?

**Answer:** There definitely is a religious connotation to the Big Bang theory. An early proponent, Georges Lemaître, was both a priest and a physicist, and I am sure that he felt this theory provided a partial scientific explanation for his deeply held spiritual beliefs. Anyway, once most researchers agreed with the Big Bang theory, they sought other specifics that could additionally strengthen their arguments.

## Other Supporting Data

One such finding is the abundance throughout the universe of hydrogen and helium. These two elements make up about 99 percent of all that is known to exist. It is theorized that hydrogen, a single proton within an electron cloud, was the initial element formed as the plasma of the primeval explosion congealed. However, as the cooling progressed, helium (the next simplest element) formed, and the ratio of hydrogen to helium (H/He, approximately 75/25) would theoretically result from this condensation effect.

It was also found that, as astronomers peered farther and farther into space, earlier galaxies (those formed many billions of years ago) appeared less mature than current ones. This made sense, as galaxies forming closer in time to the Big Bang should be younger than those found many billions of years later (after they had time to fully mature).

*Question:* So you are showing me more and more reasons to believe in the Big Bang. Do you have any doubts as to its correctness?

## Problems With Current Theory

*Answer:* There are some serious problems with this theory. The universe is thought to be flat, isotropic and homogeneous (to have the same aspect, energy, and physical characteristics) in all directions. Also, *relic* particles (specifically single-pole magnets or monopoles) hypothetically should exist but have never been found. To fit these concepts, Big Bang adherents had to interpose an extremely rapid growth—*inflation*—into the very early universe's expansion.

Thus, although it is thought that all that exists started from but one extraordinarily dense point and is continually enlarging, at a very early stage an extremely rapid expansion or inflation also occurred. It lasted for a very short time but allowed for the sameness, flatness, and absence of relic particles now observed.

Another problem is that the universe has been found to be expanding at an ever-increasing rate over the last six to ten billion years. Scientists had thought that, due to gravity's attraction (the only force believed to exist on a cosmic scale), the universe's expansion would begin to slow. The opposite was found; it appears to be growing more rapidly.

This was discovered by measuring the *real* distances of certain kinds of supernovas (all these specific supernovas—Type 1a—are of the same brightness, so one can readily determine how far away they are) and comparing these to calculated or *theoretical* distances from their redshifts (using Hubble's constant): in essence, direct verses indirect computation. When the procedures were done, disparities were uncovered; the supernovas were farther away—i.e., appeared dimmer—than theory dictated. This work was deemed important enough to secure its authors a recent Noble Prize.

**Question:** You are beginning to confuse me; perhaps you can explain yourself a little better.

**Answer:** Allow me to try; the concept is really not that difficult to understand. Let us, for example, use a redshift of 0.1, or 10 percent. This means that an interstellar body's light has shifted 10% toward the red spectrum. Thus, if on Earth the light has a frequency of 500 nanometers ($500 \times 10^{-9}$ m), when we see it coming from space it has elongated to 550 nm (a 10% increase).

Once the redshift is measured, we can determine the velocity of the object. Its speed, when based on a comparatively small redshift (10% or less), is found using a rather straightforward formula; (larger redshifts, those over 10%, due to the limiting velocity of light, would require a more complex equation). Therefore, in our example the formula is:

*Velocity = (% change in redshift) x (speed of light).*

Thus, in this case, the change in velocity is 10 percent; hence, 0.1 x 300,000 km/s = 30,000 km/s.

If we then interpose the Hubble constant (the rate or speed that the universe is expanding from us compared to its distance away—currently estimated at 67.8 km/s for every 3.26 million light-years), our object, supposedly traveling at 30,000 km/s, would be moving at about 442 times our current velocity (30,000/67.8). However, if it *were* traveling that fast, according to Big Bang theory it would be found far off in the remote void. In our example, it would be stationed 442 times as far as 3.26 million light-years, or 1.4 to 1.5 billion light-years away. (These numbers are only crude estimates, but the underlying principle is that, given a uniform increase in velocity over distance, the Hubble constant, one can determine how far away an object resides if its velocity—or red-shift—is known).

However, using the above concept, when the theoretical distance (estimated by redshift) was matched against the real distance (determined by brightness), objects appeared to be farther away than anticipated. The supernovas were fainter, more distant, than theory suggested. Thus, an expansive power (besides the initial Big Bang) was needed to account for these findings.

Since the only influence (after that of the primeval explosion) was gravity (a *contracting* force), an entirely new entity—unknown or *dark energy*—was proposed. Some cosmologists, considering it a fundamentally different form of energy, have called it *quintessence* (after the fifth element from Greek philosophy). Other theorists have used Einstein's concept of a *cosmological constant* (a fudge factor that Einstein employed to keep the universe stable, and which he later derided as his *biggest mistake*) to explain it. However, no scientist really understands its basis. Thus, dark energy and inflation have to be interposed to make the Big Bang work.

**Question:** So both inflation and unknown energy are needed to firm up the Big Bang theory. Are there any other problems?

**Answer:** Yes, another difficulty with today's theories has to do with the rapid rotation seen in galaxies, and in their much larger aggregations or clusters. Both galaxies and clusters rotate faster than their observable mass or gravity should allow. To account for this greater velocity, unknown material—*dark matter*—has also been posited to exist.

Dark matter alone is thought to have about five times as much mass as all the observable, or understandable, objects in the universe; and dark energy contains another fifteen times as much. Thus, together they comprise over 95 percent of everything, yet are essentially unknown.

*Question:* Let me see if I understand. The Big Bang, our currently acknowledged theory, made sense, as it explained Hubble's findings (the redshifts) as an expansion of the universe. If one then accepted this concept (of an ever-expanding cosmos, every spot from every other spot, with a continuously increasing velocity the farther away one looked), an initial explosion of primeval substance would be appropriate.

The background microwave radiation cemented this theory, as did the H/He ratio and the immaturity of the distant (early) galaxies. However, inflation and dark energy have to be interposed to make this explanation fit, and dark matter becomes another important stumbling block. Thus, there appear to be significant positives but disquieting negatives to our understanding.

*Answer:* You do seem to get what I am attempting to show. Let me now try to give you an alternative explanation; one using a higher dimension.

# 9

# REDSHIFT AND THE
# FOURTH DIMENSION

THE BASIC PROBLEM IS THAT 95 percent of the universe is thought to consist of unknown material—dark energy, and dark matter. We know, see, or feel less than 5 percent of everything. This is pure conjecture or fantasy.

The Earth as the center of the universe, best described by Ptolemy, was accepted gospel for almost 2000 years. It made common sense and could be easily understood. It was based on the obvious finding that, as all things "fell" to the Earth, the Earth had to be the "center" of all.

Copernicus showed that a Sun-centered concept more easily fit the facts, and Newton showed that a force—gravity—not only caused things to fall toward the Earth but also allowed for motion of the heavenly bodies. Thus, the Ptolemaic system, even though it appeared obvious and was logically consistent, was incorrect, as it was based on a faulty initial assumption.

Hubble's notion of increasing velocity with increasing distance is a similarly flawed idea. The redshift is not due to a Doppler effect; it is due to a fourth-dimensional curve. Thus, the Big Bang, based on Hubble's concept, is untenable. There is no increasing velocity; therefore, there can be no expansion.

## A Higher Dimension

We have been discussing how a lower-dimensional world would observe a higher one. We have noted that those in the lower plane would see a continuous distortion or stretching in length as they looked into the distance. I am suggesting that our three spatial directions are really superimposed on a fourth. We, of course, cannot visualize this other path, but we can see continuously increasing or distorted measurements the farther out we gaze.

Now, what astronomers note when they look farther and farther out are increasing redshifts. Hubble, in the 1920s, explained this as a Doppler effect; he surmised that objects farther away (as they had greater redshifts) were moving away faster from us. Given this explanation, the entire Big Bang, with its strengths and weaknesses, was formulated.

*If* one takes Hubble's original description of the redshifts to connote velocity, then the Big Bang is a logical theory. However, it only holds if one assumes an exceptionally rapid early inflationary phase, and an equally mysterious expansive energy source. Also, since unknown *dark* energy, and equally obscure *dark* matter, are thought to make up over 95 percent of the universe, what we really understand is a very small percentage of what exists; thus, we are fantasizing about most of reality— and this just makes no sense.

## Ptolemy vs. Copernicus

Let me give you a historical example. Prior to Copernicus, the theory of the universe was based on ancient Greek concepts best formulated by Ptolemy (a second-century astronomer). He stated that the Earth was at the center of the universe, and that the Moon, Sun, planets, and stars orbited it in perfect circles.

The theory made a lot of sense. The Earth *had* to be the center, as all things *fell* toward it. The Moon and Sun obviously moved around it, everyone could see that, and the stars, each night, also revolved about it. All these objects made flawless circles and, thus, matched the basic theory.

The planets, however, moved in strange ways—they *wandered*. To make them fit, circles within circles (epicycles) were constructed. Thus, the planets moved in perfectly round orbits with small epicycles to make them conform to a theory that had to be correct, as it coincided with observation and common sense.

*Question:* I, of course, know that Ptolemy was wrong, that the Earth is not the center of the universe, and that most things do not orbit us in so-called perfect circles. Copernicus showed that the Sun is the center, and that planets revolve about it. But why discuss this theory anyway?

*Answer:* I bring up the Ptolemaic theory for several reasons. First, it once made perfect sense to most people. The basis of its adoption was obvious. The Earth *is* the center, as all things fall to it, and the Moon, Sun and stars *do* orbit it in seemingly perfect circles. Second, if you tinkered with epicycles, most of planetary motion could also be explained.

Thus, although we know that this theory is wrong, if maintained, much of what occurs could still be explained. Copernican thinking— heliocentricity—just was better and explained findings in a simpler way. Copernicus's theory, however, was not immediately accepted. It took over 100 years, and other great minds—Kepler, Galileo, and Newton, to name a few—to finally instill the heliocentric theory into mainstream scientific thought. Therefore, an incorrect theory may seem to make the most sense, and a false concept, if tinkered with enough, can still give correct answers.

## Occam's Razor—Simplest Is Best

*Question:* Isn't there a scientific law or rule that states that the simplest theory should be the correct one?

*Answer:* Yes, there is a belief in science called "Occam's razor," which essentially notes that, for given effects or facts, the simplest explanation of their cause is the right one. Thus, the Sun as the center, with the planets revolving about it, is more straightforward and more readily conforms to findings than does a pivotal or focal Earth.

**Question:** So, given Occam's razor, where do you see problems with current Big Bang theory?

**Answer:** The basic problem is that in current theory over 95 percent of all that exists is *dark* or unknown. This is not an explanation; it is merely an *ad hoc* illusion. We see things that we do not understand; then make up exotic forms of energy–matter to explain them. A solution that allows 95 percent of reality to be imaginary is no solution at all.

**Question:** Well, the best and brightest theoretical minds seem to feel that the Big Bang is correct. Of course, there are some disbelievers, but most mainstream scientists would argue that the current concepts of inflation and dark energy are the closest fit for facts as we know them. What alternative theory can you propose?

**Answer:** I agree; the current notions (inflation and dark energy) best help to solidify Hubble's interpretation of what had been seen. But what if Hubble *misconstrued* these observations; what if the redshifts were *not* Doppler effects? Then the entire edifice would be erected on an unstable foundation. It would be just like the Ptolemaic theory of the universe, built on the assumption that we are at the center of the cosmos, since all things fall toward us. Once the concept of gravity was accepted, it explained why things fell to the Earth, but also why the Earth and other planets orbited the Sun.

## Alternate Account Of Facts

The problem with today's Big Bang theory is a fundamental one. It is not that inflation and dark energy could not account for our present idea of the universe; it is that our current concept of the universe is itself incorrect. Hubble, when he noted that redshifts increased the farther out he gazed, considered the phenomenon logically due to a Doppler effect. This had already been shown to be true for close-by stars. Their light shifted toward the red or blue spectrums depending on whether they were moving away from or toward us. Their velocity could be determined by the amount of this change.

Hubble simply assumed that the same was true as he looked farther away to more distant galaxies. Thus, the greater the redshift, the faster from us these galaxies were thought to be moving. As telescopes became more and more powerful, ever more distant entities could be observed. When the redshifts were correlated with these objects, the farther away one gazed, the greater was the shift, and the faster were these celestial bodies thought to be moving (from us and from each other).

Now, since Hubble's original explanation of Doppler shifts, newer concepts about the lengthening of the light wave (the redshift) have been given. Currently, it is not considered an alteration in length due to the velocity of objects in space, but an expansion of space itself. This enlargement causes all things to lengthen, light waves included, and carries all objects away from all others at increasing velocities. But the growth is still based on data originally described by Hubble almost 100 years ago. It has simply been refined allowing a fit for today's theories.

**Question:** I understand what you are saying. Hubble interpreted the redshift as velocity, and the farther he looked, the greater the shift, therefore the greater the velocity—hence, the Big Bang. But if his interpretation was incorrect, then the main reason for current theory, the expansion of the universe, would disappear. What do you put in its place; how do you explain the increasing redshift?

## Redshift And Fourth-Dimensional Distortion

**Answer:** We have to get back to the basic idea of a three-dimensional universe in a four-dimensional bulk. As this is almost impossible to visualize, we will simplify our universe. Instead of three spatial dimensions, we will give it one, a straight line, and place it in a two-dimensional space. It is the original Lineland concept. Let us redraw it.

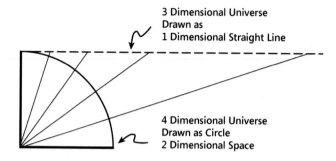

3 Dimensional Universe
Drawn as
1 Dimensional Straight Line

4 Dimensional Universe
Drawn as Circle
2 Dimensional Space

Since our three-dimensional universe (drawn as a straight line, all that we know) is really the surface of a fourth-dimensional sphere (drawn as a circle), distortions occur in all directions from any point at which we stand. In any direction, the farther out we look, the greater will similar lengths appear. Since we use light as our measuring rod, light waves are continually lengthened the farther out we gaze. Light is constantly shifted into longer or redder hues.

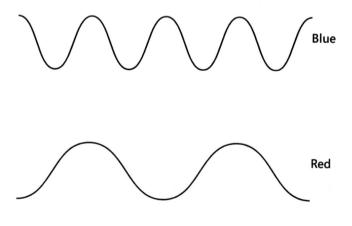

Blue

Red

Thus, what we observe in any direction is the greater the distance, the larger the shift; this, essentially, is what Hubble first saw.

Now, when Hubble initially described increasing redshifts he could see out only 50 to 100 million light-years. Today, as previously mentioned, the universe is thought to be about 13.80 billion years old; thus, Hubble only saw less than 1 percent of its entirety. However, even then, observable distortions were easily noted; the only problem was that he misinterpreted them as increasing velocity. This mistaken belief became the only explanation, and the Big Bang theory followed.

*Question:* When I look out into space on a clear, dark night I can see the Milky Way. I can even possibly discern vague nebulae. So with the naked eye one can probably see several hundred thousand light-years. Could I note significant distortions at these distances?

*Answer:* Distances observable with the unaided eye are really too insignificant to allow us to note these alterations. With the equivalent of Hubble's Mt. Wilson telescope, however, they become readily discernible. Using our current instruments, we can distinguish entities at ever-farther extremes; thus, distortions become much more apparent. We currently see celestial objects redshifted by a factor of 10, or so, equal to an expanse of just over 13 billion light-years.

## 3D Sphere (Earth)—4D Sphere (Universe)

*Question:* Our concept of three-dimensional space seems to be similar to the way we visualize the Earth, as seemingly flat. We know the Earth is a sphere, but to an individual standing in any one place it does not appear so.

*Answer:* Yes, this is a good example. We see a flat surface, but we know it is spherical. When we look into space far enough, we are visualizing the fourth-dimensional curve. Now, due to the way we are composed, we cannot understand a fourth direction. It does not exist in our three-dimensional world. However, we can note the distortions caused by that extra dimension—the redshift.

The problem is that this redshift, this lengthening of space that surrounds us, was initially described as an increase in velocity. This original explanation has held and all subsequent theories have tried to clarify it. Thus, if one assumes the redshift to be the consequence of space expanding away from all points, starting from a primeval atom, then the Big Bang is a perfectly sound description. The main problem is that it has become too cumbersome; 95 percent of all things are imaginary.

## Ptolemaic Theory And The Big Bang

*Question:* I see where you are going. Today's Big Bang theory is like the Ptolemaic construct of the universe. In both cases, the most basic assumption is wrong. In the Big Bang theory, the redshift is considered to be a continuous expansion of space (with ever-increasing velocity). In Ptolemaic thought, the Earth is considered the center of the universe (since all things fall to it).

Given basically wrong interpretations, the theories developed are consistent with them. In the Big Bang, if we consider the redshift to connote an ever-enlarging space, then an initial primordial explosion is a perfectly reasonable conclusion. Dark energy and inflation are sensible additions. In Ptolemaic theory, if we consider the Earth to be the center of the cosmos, then the circular orbits of the Moon, Sun, planets, and stars seem accurate. Of course, just as in the Big Bang, extra concepts need to be added. Ptolemaic epicycles for planetary wanderlust are today's dark energy and inflation.

*Answer:* You really seem to understand. A theory can explain, and be consistent with, findings, but if what the facts represent is initially misinterpreted then the entire theory is incorrect. The Big Bang is an explanation that has grown too convoluted to sustain; the most basic concept has been misunderstood. *There is a redshift, but there is no expansion of space.* All that the redshift signifies is the visualization, in three dimensions, of distortions caused by a fourth, but unknown direction.

# 10

## "Z" PARAMETER

T HE Z PARAMETER IS A MEASURE *of the percentage increase in the redshift. It is found using the formula:*

$$z = new - old / old.$$

*It is how the redshift is understood and employed. Once the redshift (the spectral line change) is found, and the z is calculated, a velocity for the object (galaxy or supernova) can be determined (Doppler effect), and, according to Hubble's theory, its distance from us is then known; thus, the z parameter is essentially used by astronomers to determine distance.*

### Measuring "Z"

Let me now take our new concept, three-dimensional space on a fourth-dimensional sphere, and see where it leads. The redshift is commonly measured using what is called the *z parameter*. The z parameter or value is simply a number telling us how much larger the new light wave is when compared to the old. It is essentially a percentage increase in length. For example, if the original light wave is 500 nanometers and the new wave is 1000 nm then the percentage increase is 100 percent.

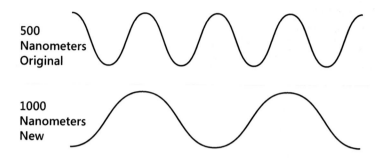

500
Nanometers
Original

1000
Nanometers
New

The math is really very simple. All one needs to do is to take the new length, subtract the old length and divide what remains by the old length. Thus, in the example given:

$$z = new - old / old$$
$$z = 1000 - 500 / 500$$
$$z = 1 \text{ or } 100\%.$$

Therefore, the change in length is one full length more than the original, or an increase of 100 percent.

**Question:** You know, I really dislike math, even math as simple as you make it. Let me try to put it into my own terms. What you are showing is merely the increase in value of something—dollars for an item, or nanometers for a light wave. Hence, if I were a street vendor normally selling Yankee T-shirts for $5.00, but on the day they won the World Series there was a rush to buy and they sold for $10.00, then, my percentage increase is the new price less the old price divided by the old price. Therefore, in my example, it is:

$$\$10.00 - \$5.00 / \$5.00 = 1.$$

This would be an increase in price of 100 percent. Is that what the z parameter tells us?

*Answer:* Your example is quite concrete but as good as any. Thus, the $z$ parameter or value is simply the percentage increase (written as a whole number) due to the redshift. Therefore, a $z$ of 0.01 equals an increase of 1 percent; a $z$ of 0.1, 10 percent; a $z$ of 0.5, 50 percent; and a $z$ of 1.0 amounts to an increase of 100 percent, or a doubling in length. As we look ever farther out we see redshifts with $z$ values up to 10 (1,000% increase), or lengths that are now 11 times as great as the original.

*Question:* Okay, the $z$ parameter appears to be straightforward. Even I understand the math. The equation is quite simple ($z$ = new - old / old). So why is this concept so important?

## "Z"—Velocity And Distance

*Answer:* When Hubble first noted the stretching that occurred with distant light waves (the continually enlarging redshift), he assumed that this would persist indefinitely. The expansion of the universe (the Big Bang) was then proposed; and this continuous growth was encoded as Hubble's law. At the present time, Hubble's constant (the velocity of an object compared to its distance) is estimated at 67.8 km/s for every 3.26 million light-years. Thus, if one knows the speed of an object, one can determine how far away it is.

Remember the example we discussed before: If something is moving at 30,000 km/s, then it would be between 1.4 and 1.5 billion light-years away. Therefore, given the Hubble constant, it is thought that, once a velocity is computed, a distance can be found.

*Question:* But we were discussing the $z$ parameter, the percentage increase in the wavelength of light. How do we get a velocity to fit into Hubble's law from this $z$ value?

*Answer:* Let me again show you some formulas that allow for this conversion. If the $z$ value is small, equal to or less than 0.1 (or 10%), then the equation, as already mentioned, is easy to understand:

$$z \ (times) \ c = v$$

where $c$ is the speed of light (300,000 km/s) and $v$ is the object's velocity. If we divide both sides by $c$:

$$z = v \ / \ c.$$

If we know the $z$ of an object, we then know its velocity. For example, if $z = 0.1$, then:

$$0.1 = (v) \ / \ 300,000 \ km/s, \ thus,$$
$$0.1 = (30,000 \ km/s) \ / \ 300,000 \ km/s; \ hence,$$
$$v = 30,000 \ km/s.$$

**Question:** That *is* really straightforward. What happens when $z$ is greater than 0.1?

**Answer:** We now have to use a formula that takes into account the limiting velocity of light (as nothing can exceed that speed). It looks complex but is not really that difficult to use:

$$z = square \ root \ of \ (1 + v/c \ / \ 1 - v/c) -1.$$

Thus, for example, a $z$ of 0.50 yields a velocity of about 115,000 km/s, not 150,000 km/s, as would be expected with the simpler equation. The higher the $z$, the greater the velocity; however, it can only continuously *approach* the speed of light, it can never *reach* it.

**Question:** Okay, I see where you are going. $Z$ is an easy concept to understand. It is essentially the percentage increase in anything; astronomers use it to determine the changes in light coming from a distant source. I guess it is measured with a spectroscope or similar instrument and compared to the same light as seen on the Earth. Then, using the above formulas, it is transformed into a velocity. Once it is a velocity, using Hubble's law of expansion it can be used to find a distance. Therefore, every $z$ value really represents a distance away from us.

*Answer:* You really do understand. If we assume an expanding universe, with ever-increasing velocities, using the $z$ values to determine speeds allows us to estimate distances. The problem is that, when the distances computed by use of the $z$ values are compared to the *actual* distances found using supernova intensities, there are discrepancies. The $z$ values come up with distances that are less than those noted by the intrinsic brightness of these exploding stars (i.e., the supernovas are farther away than theory predicts). Thus, the universe is thought to be expanding faster than was originally expected (from the Big Bang alone), and an expansive force (dark or unknown energy) is interposed as the cause.

*Question:* Let me try to summarize what you have been saying. We have $z$ values by spectroscopic determination, and, hence, we have velocities by use of the above equations. Given these velocities, we can then compute distances, but these distances come up short. The *real* distances, measured by direct observation (intrinsic brightness), are greater than those computed by Hubble's constant. Therefore, an expansion is presumed, and a force for this expansion (dark energy) is proposed. Is that the problem?

*Answer:* That is precisely the problem. To solve it we must, once again, return to our concept of a higher dimension.

# 11

# TANGENTS AND "Z"

I N A RIGHT TRIANGLE, *the tangent is defined as the opposite side divided by the adjacent side. In our simplified model of the universe (straight line equals 3D, and circle equates with 4D curve) the tangent's opposite side is the ever-increasing segment (as one looks farther and farther into space), and its adjacent side is the unchanging radius of the circle.*

*As the angle increases (to just under $90^0$) the tangent begins to enlarge exponentially; thus, the opposite side (the varying distance) continuously expands, and z (the ratio: new [opposite] - old [arc] / old [arc]) likewise rapidly grows. Therefore, the expansion we find as we peer ever deeper into space (the constant enlargement of the z parameter) does not constitute a velocity increase; it simply signifies a higher-dimensional curve.*

## Right Triangles And Tangents

**Answer:** I am sure you know what a triangle is.

**Question:** Of course I do; it has three sides, and its angles add up to $180^0$. Why do you even ask?

**Answer:** Don't be upset, I only asked to get some basics down. Now, there are different types of triangles. A *right* triangle has one angle of $90^0$;

thus the other two angles add up to an additional 90⁰, or 180⁰ altogether. The side opposite the right angle is called the *hypotenuse*. Let me draw one. Each angle has a letter, and each line is denoted by two letters. Thus, angle B is $90^0$ and side AC is the hypotenuse.

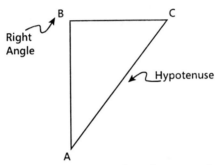

Now there are ways of measuring lengths in a right triangle if we know the angle and only one of the three sides. By using sines, cosines, or tangents, we can determine unknown lengths when aware of only one side. Do you remember the mnemonic SOH CAH TOA?

**Question:** It sounds like an Indian war chant. No, I do not recall this mnemonic.

**Answer:** It means that sine is opposite divided by hypotenuse, cosine adjacent by hypotenuse, and tangent opposite by adjacent. We are just going to use the tangent (or opposite divided by adjacent). Now, given any angle from a fraction of $1^0$ to almost $90^0$, a corresponding tangent is assigned. A simple example is $45^0$.

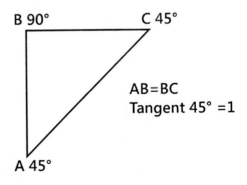

If angle A equals $45^0$, then angle C also equals $45^0$ as the two angles must equal $90^0$. Thus, the opposite and adjacent sides BC and AB are equal and BC/AB equals 1. Therefore, the tangent of $45^0$ is 1. Some other tangents are as follows:

$$1^0 = .017$$
$$30^0 = .577$$
$$60^0 = 1.732$$
$$89^0 = 13.6$$
$$89\frac{1}{2}^0 = 114.6.$$

*Question:* Why are you now giving me all this useless material? What do you really expect me to do with this information overload?

## Tangents Define "Z"

*Answer:* I want to give you a feel for some tangents, how they increase as the angles get larger. Let me simplify my concept again. Remember, we are drawing our three-dimensional world as a one-dimensional straight line, and the four-dimensional sphere as a two-dimensional circle. We are then extrapolating equal sizes on the circumference of the circle to what they appear to be on the straight line. This is done by use of tangents. Let me show you.

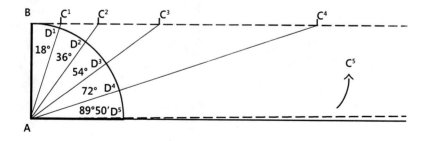

We are drawing a right triangle ABC. AB is the radius of the circle. BC is the length of the line in the one-dimensional universe (our simplified three-dimensional world). Finally, BD is the length of the segment of the circle (our simplified four-dimensional sphere). Since we are cutting the segments of the circle into five equal lengths, and since we are using only one-fourth of the entire circle (or 90°) each segment has an angle of 90°/5, or 18°. Thus, we are looking at tangents of 18°, 36°, 54°, 72°, and finally just less than 90°—lets us say 89°50' or 89.83°. The tangents (found in any standard table) are as follows:

$$18° = 0.32492$$
$$36° = 0.72654$$
$$54° = 1.37638$$
$$72° = 3.07768$$
$$89°50' = 343.774.$$

**Question:** I see what you are doing. Each segment on the quarter circle is the same length. Since there are five segments, each segment is a fifth of the total. But each segment on the straight line keeps increasing in length, and as you go toward 90° on the circle the length gets very great indeed.

**Answer:** That is correct. Remember, a tangent equals opposite/adjacent, and the one-fourth of the circle that we drew is the entire universe (to our one-dimensional line). Nothing exists beyond 90°; it signifies infinity (the tangent of 90°). Let us now do the math; it is quite straightforward.

The length of BC[1] is based on the tangent of 18°. This equals BC[1]/AB. So $0.32492 = BC^1/AB$. Now AB is the radius of the circle, which is the same for all the triangles. Since we are saying the segment of the circle—BD[5]—is the entire universe (approximately 13.80 billion light-years), then the complete circle would be 4 times 13.80, or about 55.2 billion light-years. The radius is found using the formula:

$$Circumference = 2\pi r$$
$$circumference/2\pi = r$$

$$\pi = 3.14159, thus,$$
$$2\pi = 6.28318, hence,$$
$$55.2/6.28318 = 8.785.$$

Therefore, the radius of our circle is, more or less, 8.785 billion light-years. Now, remember, given the tangent and the adjacent side we can find the opposite side. Thus, the math is relatively simple:

tangent = opposite/adjacent
adjacent = 8.785
tangent = opposite/8.785, hence,
tangent x (8.785) = opposite;
or, the length in our world.

If we plug in the numbers we get the following lengths:

$BC^1 = 0.32492 \times 8.785 = 2.85$
$BC^2 = 0.72654 \times 8.785 = 6.38$
$BC^3 = 1.37638 \times 8.785 = 12.09$
$BC^4 = 3.07768 \times 8.785 = 27.04$
$BC^5 = 343.774 \times 8.785 = 3020.$

Thus, the length (as seen in our universe) gets much greater as we delve ever farther into the next dimension.

Now remember, z is the change in length of the elongated (new) vs. the original (old) light wave, and the original wave is equivalent to the length of the segment of the circle. As we have conveniently cut the circle into five equal parts, each segment equals 13.80/5, or 2.76 billion light-years. So:

$BD^1 = 2.76$
$BD^2 = 5.52$
$BD^3 = 8.28$
$BD^4 = 11.04$
$BD^5 = 13.80$ (almost).

Thus, the z values (new - old / old) are as follows:

$$Z^1 = BC^1 - BD^1/BD^1 = 2.85 - 2.76/2.76 = 0.09/2.76 = 0.033$$
$$Z^2 = BC^2 - BD^2/BD^2 = 6.38 - 5.52/5.52 = 0.86/5.52 = 0.156$$
$$Z^3 = BC^3 - BD^3/BD^3 = 12.09 - 8.28/8.28 = 3.81/8.28 = 0.46$$
$$Z^4 = BC^4 - BD^4/BD^4 = 27.04 - 11.04/11.04 = 16.0/11.04 = 1.45$$
$$Z^5 = BC^5 - BD^5/BD^5 = 3020 - 13.80/13.80 = 3006/13.80 = 217.8.$$

Finally, if we arrange it as light-years vs. z we get:

| light-years (billions) | z |
|---|---|
| 2.76 | 0.033 |
| 5.52 | 0.156 |
| 8.28 | 0.46 |
| 11.04 | 1.45 |
| 13.80 (almost) | 217.8. |

## "Z" Parameter Is The Fourth Dimension

**Question:** I am glad that you felt the math was straightforward. It was a little too dense for my taste. However, I think you are saying that, if we assume the real universe to be a fourth-dimensional sphere, but that our comprehension is only flat three-dimensional space, then, as we gaze outward, ever farther, light from the most distant objects will be redshifted by the amounts shown.

**Answer:** I am sorry that you found the previous part bothersome. The math really was not difficult, but perhaps somewhat laborious. However, you do appear to grasp what I am attempting to show.

Astronomers can easily measure z values, determine, by the Doppler effect, supposed velocities, and, via Hubble's constant, ascertain distances. However, z values really just measure a higher-dimensional curve; they do not represent velocities. Therefore, once we find them (by spectroscopy), we can then calculate the distortions in our plane and establish true distances. Since the redshift as defined by the $z$ parameter simply represents the fourth-dimensional curve, obviously, as we look farther and farther out on the curve, objects appear ever more warped (or redshifted). But they are not moving away from us at continuously escalating velocities; there is no expanding space starting from a central point. The increasing redshift does not mean there was a Big Bang. It simply means that, as we gaze ever farther into the next dimension, lengths keep distorting, and the greater the distance, the greater the change.

# 12

# DARK ENERGY

ARK ENERGY IS CONSIDERED *the cause of a supposed "increasing" expansion of the universe. It is found when comparing supernova data with distances based on Hubble's constant. However, since there was no initial expansion and, therefore, Hubble's constant is meaningless, what, then, really, is dark energy?*

*It becomes the difference noted (in z value-determined distances) when employing tangents (a fourth-dimensional curve) or using Hubble's constant (a supposed expansion of the universe). It is an illusion triggered by a fundamental misunderstanding— the redshift is not a Doppler effect but is simply due to a higher-dimensional curve.*

*Therefore, the difference—supposedly caused by dark energy—is nonexistent; it is nothing at all. It disappears along with inflation, once Hubble's theory is discarded.*

## Does Dark Energy Exist?

**Question:** So where does that leave us with the whole concept of dark energy or an increasing expansion of the universe?

**Answer:** Obviously, if there is *no* expansion, then there is no *increasing* expansion (or dark energy). As we already stated, the $z$ values are generally thought to represent velocity, and since the universe is thought to expand according to Hubble's law (the greater the distance, the greater the speed), Hubble's constant is used to determine an object's remoteness. Therefore, every z supposedly tells us a distance, and we could now chart distance against these values.

However, when this is done and (using supernova data) the actual expanse is measured, it has been found to be greater than estimates generated by Hubble's law. This increase (in distance) is thought to be due to an unknown force (dark energy) working against gravity, which has expanded the universe over the last 10 to 12 billion years.

If, instead, one were to use z values based on a higher-dimensional curve (the tangents described), then the distances computed for each z value, for almost the last 12, or so, billion years, would be greater than are commonly projected. Thus, the estimates that employ Hubble's constant are incorrect—not because there is a *dark* force of increasing expansion, but because the original expansion itself *never occurred*.

Let me chart this based on our concept of the entire universe being equal to just under $90^0$ (or a quarter of a circle). Thus, if 13.80 billion light-years equals almost $90^0$, then $1^0$ equals approximately 0.153 billion, or 153 million light-years (13.80/90). Our table now includes columns for angles, tangents for those angles, distances in billions of light-years, the resultant z values determined by those tangents, velocities estimated by those distances (via Hubble's law), and, finally, the z values calculated from those velocities.

| Angle | Tangent | Distance ($10^9$ Lys) | Z (Tangent) | Velocity (km/s) | Z (Hubble) |
|-------|---------|-----------|-------------|-----------------|------------|
| $1^0$ | 0.0175 | 0.153 | 0.0003 | 3200 | 0.01 |
| $5^0$ | 0.0875 | 0.767 | 0.002 | 16000 | 0.05 |
| $10^0$ | 0.176 | 1.533 | 0.008 | 32000 | 0.10 |
| $20^0$ | 0.364 | 3.067 | 0.043 | 64000 | 0.23 |
| $30^0$ | 0.577 | 4.60 | 0.1 | 96000 | 0.4 |
| $40^0$ | 0.839 | 6.13 | 0.2 | 127000 | 0.6 |
| $50^0$ | 1.19 | 7.66 | 0.36 | 159000 | 0.8 |
| $60^0$ | 1.73 | 9.20 | 0.65 | 191000 | 1.1 |
| $70^0$ | 2.75 | 10.73 | 1.25 | 223000 | 1.6 |
| $75^0$ | 3.73 | 11.50 | 1.85 | 239000 | 2.0 |
| $80^0$ | 5.67 | 12.26 | 3.06 | 255000 | 2.5 |

## If No Dark Energy, Then No Big Bang

**Question:** You know I do not like math, even simple math, and your table is getting too complicated for me. I understand the first column. It just gives angles from $1^0$ to close to $90^0$. The next column is merely the tangent of that angle; I guess you looked it up in a book. The third column is the distance where each degree equals about 153 million light-years. The $Z$(Tangent) column is what you have been discussing for a while now. It is the equation ($z = $ new - old / old), where the new (distance) is the tangent (the straight line segment) determined using the radius of the circle. But where did you get the last two columns—Velocity and $Z$(Hubble)?

**Answer:** The fifth column is simply based on Hubble's constant of expansion of 67.8 km/s for every 3.26 million light-years; the known distance (column 3) is divided by 3.26 million to get the number of times it has increased and then multiplied by 67.8 km/s to get the supposed velocity. The last column is a determination of the $z$ parameter based on this velocity. To get $z$ using the velocity, one has to use the previously discussed formula, which takes into account the limiting velocity of the speed of light:

$$Z = square\ root\ of\ (1 + v/c\ /\ 1 - v/c) - 1.$$

It looks complicated, but is not really hard to use. Thus, the final column, $Z$(Hubble), is based on distance being proportional to increasing velocity, or to an expansion of space—Hubble's law.

**Question:** Okay, I think I understand. What you apparently have done is to use a known distance to derive $z$ values in two distinct ways (by fourth-dimensional tangents, and by Hubble's law/constant). The tangent method allows for a greater distance at every $z$ up to just over a value of 2. For instance, a $z$ of 0.1 via Hubble is about 1.5 billion light-years, whereas the same $z$ via tangent would be approximately 4.6 billion.

Therefore, whenever an object is noted, and its redshift (or $z$ value) is

determined by spectroscopy and then (via Hubble's constant) used to calculate a velocity, which, in turn, gives us a distance, this distance has been
inaccurate for close to the last 12 billion years. Hubble's supposed law of
expansion places the object nearer than it is shown to be by supernova data.
Is this the reason for the controversy that led to the concept of dark energy?

*Answer:* Yes, it is. Astronomers found celestial bodies (supernovas),
determined their redshifts, calculated their z-value distances, and then
noted their real distances using intrinsic brightness. These real distances
were always greater than Hubble's law allowed. Thus, for almost 12 or
so billion years, space appears to have been expanding faster than initially
thought; objects are really farther away than Hubble's law places them,
and a mysterious force has been attributed to this expansion to explain
the discrepancy—dark energy.

Let me give you one more example with a table that should make it
easier to understand; here z is found (by spectroscopy), and distance is
determined both by Hubble's expansion and by a fourth-dimensional
curve using tangents—the difference is the supposed dark energy.

| $Z$ | Distance ($10^9$ Lys) Hubble/Expansion | Distance ($10^9$ Lys) 4D curve/tangent | Difference Dark Energy? |
|-----|------------------|------------------|------------------|
| 0.1 | 1.5 | 4.6 | +3.1 |
| 0.2 | 2.8 | 6.1 | +3.3 |
| 0.4 | 4.6 | 8.0 | +3.4 |
| 0.8 | 7.7 | 9.7 | +2.0 |
| 1.2 | 9.5 | 10.7 | +1.2 |
| 2.0 | 11.5 | 11.6 | +0.1 |
| 2.5 | 12.3 | 11.9 | −0.4 |

Therefore, once a z value is calculated (from the redshift found via
spectroscopy—column 1), using a 4D curve/tangent (column 3) places
that object, more appropriately, farther away than using the concept of
Hubble/expansion (column 2). The difference is the reputed expansion of

dark energy (column 4), and it is present to almost 12 billion light-years.

Thus, there is no expansion in the first place (the usual explanation for Hubble's findings is mistaken) there is just a higher-dimensional curve that causes a stretching in the fabric of space (more pronounced the farther out one looks). The recently discovered *increasing* expansion—dark energy—does not exist; it is a chimera, an illusion.

Let me show you this concept more clearly, using the tables discussed in the form of a graph (with z as the horizontal axis and distance as the vertical axis). What you will see is a greater distance for all z values (until just around 2.0) when comparing a fourth-dimensional curve to one based on Hubble's expansion. The presumptive dark energy is simply the difference between the two curves.

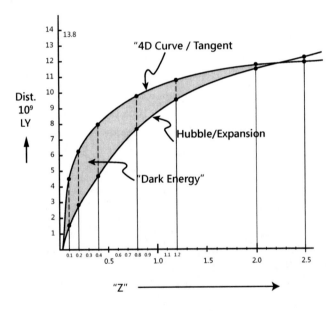

At All Z Values up until just above 2.0
4D curve distance greater than Hubble

You see, astronomers used the z of objects (based on the redshifting of their spectral lines) as a shortcut to determine their remoteness in space; but the distances calculated were mistaken, as they were founded on an incorrect assumption (Hubble's constant of expansion). These computed values (up to almost 12 billion light-years) were always closer than the real distances (determined by intrinsic brightness). If they had used a table based on a higher-dimensional curve the $z$ values would give more accurate estimates of the real (intrinsic brightness) distances and no dark energy would be necessary.

*Question:* I think I understand. The $z$ parameters estimated from observing redshifts are used to determine distances. These distances are based on Hubble's constant of expansion, but, since there *is* no expansion, Hubble's constant is invalid; one gets distances that are wrong, that are too small. Therefore, we do not need dark energy to explain *increased* expansion, as there was *no* expansion—no Big Bang—in the first place.

So, once you do away with the Big Bang, you also do away with the need for inflation and the phenomenon of dark energy. But what about other concepts that helped to establish the Big Bang theory? What about cosmic microwave background radiation, the H/He ratio, distant immature galaxies, and even dark matter? How can you account for all of these?

*Answer:* Let us expand our thinking a bit by simplifying our model. We can start, once again, by drawing our known three-dimensional universe as a one-dimensional straight line, and its four-dimensional bend as a two-dimensional circle. It should help us to more easily comprehend what I am trying to explain.

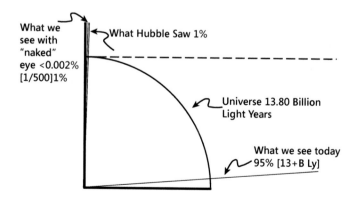

We can see from the drawing that, with the naked eye, we view but a tiny fraction of the universe; only several hundred thousand light-years, or less than 1/50,000 (1/500 of 1%) of the entirety. Hubble, with the Mt. Wilson telescope, perceived less than 1 percent as well; currently, with our ever-more-powerful instruments, we distinguish up to 95 percent (of the total distance) or 13-plus billion light-years. Thus, today we are noting greater and greater redshifts. What do you think will happen as we keep looking ever farther outward?

# 13

# VERY SMALL—VERY LARGE

$A$ *S WE GAZE OUTWARD EVER FARTHER, we finally come to a place after which nothing can be seen. At this point we are just less than $90^0$ on the fourth-dimensional curve, to be concise, one Planck length away from $90^0$. The z parameter at this juncture is extremely large (about $10^{32}$), and light waves are now so stretched that they are about the same size as the entire universe (1.3 x $10^{26}$ meters). Once greater than that, the universe no longer is visible—it disappears.*

*Cosmic microwave background radiation, thought by most mainstream scientists to be due to an expanding universe, can be accounted for by the tremendous stretching that occurs at this distance. Thus, the last possible visible part of the universe (Planck length, about $10^{-35}$ m) is extended by the z factor (approximately $10^{32}$), and the result is microwave radiation ($10^{-35}$ m x $10^{32}$ = $10^{-3}$ m).*

## Higher And Higher "Z" Values

**Question:** When I look at your z values, I see a marked, almost exponential increase starting at around 11 billion light-years; at that distance z is approximately 1.5 or, as you have stated, the light waves have been lengthened or redshifted around 2.5 times. When using your calculations, if I were to peer out to almost 13.80 billion light-years, I would see z values of over 200. So in the farthest, most distant reaches of the cosmos, the $z$ parameters have greatly expanded.

**Answer:** Let me show you how the $z$ values start to dramatically increase as we go farther and farther from our spot in the universe. Let us go from an angle of $85^0$, equal to 13.033 billion light-years, to that of $89^0$, or 13.647 billion light-years, and then to $89^050'$ or 13.774 billion light-years.

| Angle | Distance (Billion Lys) | Tangent | Z |
|-------|------------------------|---------|---|
| $85^0$ | 13.033 | 11.43 | 6.71 |
| $86^0$ | 13.187 | 14.30 | 8.53 |
| $87^0$ | 13.340 | 19.08 | 11.57 |
| $88^0$ | 13.493 | 28.64 | 17.65 |
| $89^0$ | 13.647 | 57.29 | 35.88 |
| $89^050'$ | 13.774 | 343.77 | 218 (approx.) |

If we go from $85^0$ to $89^0$, $z$ goes from 6.71 to 35.88, a five- to six-fold increase. Then, if we go from $89^0$ to $89^050'$ $z$ goes up by another factor of six or so. Thus, starting from a distance of 13.774 billion light-years, the light now reaching us has been shifted by just over 200 times toward and beyond the red spectrum.

If the universe supposedly began 13.80 billion years ago, 13.774 billion years in the past would be 0.026 billion years (13.80 - 13.774) after the start of the universe. Thus, from just slightly more than 25 million years after the universe is thought to have begun, light has shifted from its normal range by a factor of over 200 times toward, and through, the red spectrum. If we assume normal light has an average wavelength of 550 nanometers, or $550 \times 10^{-9}$ meters, then 200 times this is $110,000 \times 10^{-9}$ m, or about $10^{-4}$ m. This is almost in the microwave range ($10^{-3}$ to $10^{-2}$ m).

**Question:** So what occurs if we keep going ever farther back in time; what happens to this redshift?

## Limits—Planck Time, Planck Distance

*Answer:* Let us look back. We generally assume that, from 380,000 years after the so-called Big Bang to the present, the redshift has increased about 1100 times. Thus, light which originally was $550 \times 10^{-9}$ m (380,000 years after the Big Bang) now would be about $605,000 \times 10^{-9}$ m, or approximately $6 \times 10^{-4}$ m. This is the epoch of the cosmic microwave background radiation.

We also know that, at around 10 seconds after the supposed Big Bang, $z$ was about 1 billion, and that, at 1 second (9 seconds earlier), $z$ was more or less 3 billion. If we keep on looking back, making an angle closer and closer to $90^0$, we reach a point called the *Planck time*. This is the smallest time interval allowed (or $5.4 \times 10^{-44}$ of a second). Since light travels at $3 \times 10^8$ m/s, this is $1.6 \times 10^{-35}$ of a meter. The angle now is 89.99..9 to the $59^{th}$ power or, if we were to write it: $89.99,999,999,999,999,999,999,999,999,999,999,999,999,$ $999,999,999,999,999,999^0$. Smaller than this time interval, nothing exists in our universe.

*Question:* Why is there a limit as to time and size in our universe? Why can't something last less than $5.4 \times 10^{-44}$ of a second or be smaller than $1.6 \times 10^{-35}$ of a meter?

*Answer:* This was worked out by Max Planck around 1900. He took the basic constants of nature—gravity, the speed of light, and what was later named, in his honor, the *Planck constant*—and derived these numbers. However, we have seen that the $z$ value keeps increasing. Since it is a measure of a wavelength of light, once that wavelength stretches beyond the size of our universe, our universe becomes invisible: It no longer exists.

Therefore, as the $z$ value ever more rapidly expands (the bigger our angle becomes), at an angle of greater than $89.99..9^{59}$ $z$ becomes so large that light waves extend farther than $10^{26}$ m, the size of our universe. Our world, if smaller than a wavelength of visible light, no longer can be seen and disappears.

*Question:* So are you saying that measuring $z$ leads to the same finding as Planck discovered with his universal scale? Are you saying that there is a natural limit to size, and that things can only get so small, for, at some point, what we know as real disappears?

*Answer:* That is what I am saying. At just less than the Planck time ($5.4 \times 10^{-44}$ s), the normal wavelength of light has stretched to over $10^{26}$ m and is greater than the universe. The universe, as it is no longer visible, no longer exists.

The way I got to the $z$ value was by the use of tangents. However, since I did not have one that fit an angle of that size, I kept factoring a number, found by *hit or miss*, that met existing criteria ($z$ more or less equal to 1080 at 380,000 years, 1 billion at 10 seconds, and 3 billion at 1 second after the supposed Big Bang).

Using this factor, $z$ came out to a little more than $10^{32}$. If an average light wave is around $550 \times 10^{-9}$ m, then, if we multiply by about $10^{32}$, we get in the range of $5.5 \times 10^{25}$ m, which is just less than the size of the universe ($1.3 \times 10^{26}$ m). Therefore, a wavelength of light at the Planck length/time is almost as large as the universe; greater than that, the universe, as no longer visible, would disappear.

*Question:* Okay, I see where your concepts are leading. If we talk of the universe as 13.80 billion years old, and if we keep getting a smaller and smaller fraction of this, we finally get to $5.4 \times 10^{-44}$ s; after this interval, nothing exists. So if we were to draw it out it should look like the following diagram (presented in its entirety on the next page).

*Answer:* Yes, you do see where I am going. The universe is thought to be 13.80 billion years old. Light from that time has been traveling over a distance of $1.3 \times 10^{26}$ m. But the wavelength has been stretched during that period from visible light (about $550 \times 10^{-9}$ m) to almost $10^{26}$ m, so, beyond that point, nothing else exists.

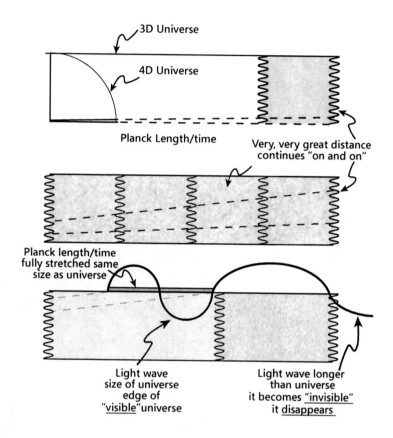

## CMBR: An Alternate Explanation

Let me discuss another interesting concept. Remember, the main reason the Big Bang became the accepted theory of the universe was the finding of the cosmic microwave background radiation (CMBR) in 1964 by Penzias and Wilson using their radio telescope at Bell Labs. The *noise* that could not be erased was found to be exactly what theorists had predicted to occur about 380,000 years after the initial Big Bang. At that time, the fog of plasma would have cooled into the clarity of atoms, and the light of the initial explosion would have become visible. As the universe would have since expanded by over 1000 times, this light would have stretched into the infrared and finally microwave spectrums. Thus, the *light* of the initial Big Bang is today's CMBR.

However, if the Big Bang did not occur, if the universe did not originate from a primeval *atom*, then this scenario is incorrect. Remember, at just less than Planck time, when we have an angle just greater than $89.99..9^{59}$ degrees, the $z$ is so large that normal light waves are longer than the universe, and it disappears.

Now, the value of $z$, as already shown, in order to stretch visible light from $550 \times 10^{-9}$ m to about $10^{26}$ m, needs to be around $10^{32}$. The last possible distance viewable would be one Planck length less than $90^0$ ($1.6 \times 10^{-35}$ m). If it were stretched by $10^{32}$, we would get the equivalent of radiation of $10^{-3}$ m ($10^{-35}$ m $\times 10^{32}$). This is our classic microwave background noise.

Thus, another explanation for CMBR (the cornerstone of the Big Bang theory) does not use the remnants of some initial *explosion* subsequently expanded over 1000 times, but is simply the background noise of our universe stretched from one Planck interval (just before it disappears) to current microwave length. The noise is that of the very edge of our world.

## Other Difficult Facts

**Question:** I see where you are going. You've said that, if we accept your idea of a fourth-dimensional curve as the real universe, we could ignore inflation and dark energy. Now we can do away with CMBR as emanating from a Big Bang. That still leaves the H/He ratio, and distant immature appearing galaxies. It also does not explain dark matter.

**Answer:** I cannot give you a good reason for the H/He ratio being what it is; but remember, the universe *to us* cannot exceed 13.80 billion years, whereas the real universe's age is unknown. If we are on a rotating fourth-dimensional sphere moving at the speed of light, our visible universe is always 13.80 billion light-years. Everything that is known to us exists in that time, but the universe has been around for eons, perhaps for an infinite time. So there may be a steady rate of hydrogen to helium conversion where hydrogen remains at 75 percent.

However, the currently accepted rationale for the H/He ratio (75/25) is based on rampant speculation of what supposedly occurred in the first three minutes after the Big Bang—an unimaginably hot and energetic period. It assumes a photon/proton proportion of $10^9/1$ to allow for the early conversion of hydrogen to helium (leading to the relationship now found). But the quantity of photons is based on a Big Bang–expansionary scenario that never existed, and the proton estimate may also be unreliable (by a factor of two to three).

Even then, there still are problems with the current concept—specifically, the lithium anomaly (less is present than theory dictates) and the deuterium abundance (more exists than can be accounted for without an imaginative addition of dark matter). Although I cannot give you a specific rationale, if there was no exploding primeval atom, then the reasons given by Big Bang theory must be incorrect.

Concerning the supposed immaturity of distant, early galaxies, this is a difficult evaluation to make due to their remoteness and inherent vagueness. Recent evaluations of these entities using gravity lensing and background quasar illumination show unexpected aging—metallicity, or the presence of complex elements—in what should be juvenile galaxies. Thus, even at great distances (in excess of 12 billion light-years), there is a maturity that abrogates current theory. Hopefully the *James Webb Space Telescope* (a much more powerful portal to our heavens), planned for launching somewhere around 2018, will rectify this mistaken concept of distance and youth. As to dark matter, I will try to get to that later in our discussion.

However, the most important fact was always CMBR, and the enormous stretching that occurs just as one reaches the Planck length is a simpler explanation for the background noise. Just getting rid of dark energy is a great relief, and the early, almost instantaneous inflationary event required to smooth out the universe was always too ingenious to really satisfy me. I am not alone in feeling that inflation may be mistaken. There is an intriguing article in a recent *Scientific American* publication by Paul Steinhardt, an early proponent of inflation, in which is noted that the odds of inflation having occurred are about 1 to $10^{(googol)}$, a number so large that it is impossible to write. This is because it would entail going from a high entropy, or disordered state, to a low one—as if a cup of coffee fell off a table, broke into pieces, and spilled its contents, then miraculously reappeared on the table complete with all its liquid intact.

Thus, if there is no real possibility of inflation, then the entire Big Bang premise is wrong. If dark energy is an illusion, and CMBR is more simply explained by the enormous stretch at the edge of our universe, then a fourth-dimensional curve is much more reasonable than an expanding Big Bang.

Noble Prizes were given both for the discovery of CMBR and for the recent supernova findings thought to reveal an accelerated expansion with dark energy; both prizes were well deserved, but not for the reasons presumed. These discoveries did not establish a Big Bang theory, they instead disproved one. Therefore, given this fourth-dimensional curve, let us try to understand how it leads to the physical world we see and feel and our place within it.

# Part Two

## THE PRIMARY EFFECT:
## THE PHYSICAL WORLD

THE SECOND PART OF THIS BOOK *explores the effect of a higher dimension in the physical world. It is seen in the redshift of the cosmos; it is real and is the basis of all that exists. Using this concept, gravity is shown to be its manifestation—a universal force felt instantaneously. Protons represent other worlds and all are interconnected; all are similar 4D spheres with 3D surfaces "rotating" at the speed of light.*

*Entanglement of all universes is described. Given a 4D perspective, quantum weirdness (with "spooky" action at a distance and wave–particle duality) becomes understandable. Electricity and magnetism are seen as manifestations of this force and Einstein's special theory of relativity is but an outgrowth of how this force is appreciated, in third- and fourth-dimensional terms.*

# SECTION III

## *Force, Distance, And Time*

FOURTH-DIMENSIONAL DISTANCE (D) IS UNDERSTOOD *by us as time (T). This is because the universe moves constantly toward the fourth dimension at the speed of light, "c"; thus,* $c = D/T$, *or* $(c)T = D$. *Since* c *is a constant, T is essentially the same as D, or, one can say, time and distance are the same.*

*In each Planck instant we enter the future (the 4D), and a new 3D universe is re-established; therefore, the velocity of gravitational force—the essence that forms and holds our world together—must travel the whole extent of our universe every such moment. It must be instantaneous.*

*Individual protons represent entire universes similar to ours. The strong nuclear force (SNF—holding protons and nuclei together), and gravity, therefore, are comparable; both can be understood as the centripetal force necessary to keep mass spinning toward a higher dimension at* c. *They are both equal to* $mc^2$ *(the energy of an entire universe) directed inwardly—toward the 4D.*

*Finally, there is a universal ratio—the Fine Structure Constant (1/137, or about* $10^{-2}$*)—which can be understood as the ratio of the electromagnetic force (EMF) to SNF. Therefore, if SNF appears* $10^{41}$ *times as potent as gravity, EMF would appear* $10^{39}$ *times as great* $(10^{39}/10^{41} = 10^{-2})$. *However, these tremendous discrepancies are tied into measurement in different dimensions. In reality, all these entities are the same* $(mc^2)$.

# 14

# TIME

I F ONE CONSIDERS THE UNIVERSE to be a 3D surface rotating at the speed of light toward the 4D on a great higher-dimensional sphere, as each new moment comes into existence an old one must disappear—the universe can never exceed $90^0$. However, since the direction toward which we travel cannot be conceived of as a distance (there are but three spatial dimensions in our understandable world), we visualize it as time.

An easy analogy is thinking of someone who is 40 miles distant, and traveling at 40 mph, as either 40 miles or 1 hour away. Thus, distance can be understood as time if there is a constant speed involved. In the universe, this constant is the speed of light.

The smallest unit of time is the Planck instant ($5.4 \times 10^{-44}$ s). Since the 3D universe reconstitutes itself each such moment, the force of this recombination (gravity) must be felt over the entire universe ($1.3 \times 10^{26}$ m) each time, and, thus, its speed must be "instantaneous" (or $2.4 \times 10^{69}$ m/s).

## Instant-Planck Time

*Answer:* Whenever we peer out into space, we are looking back into time. In fact, whenever we see anything at all, we are viewing the past. Nothing that we see or feel, not even our thoughts, takes place at this exact moment. All things have already occurred.

*Question:* That makes no sense. Of course there is a current instant. It may not be long, but it does exist.

*Answer:* You are right. There is a current moment. It lasts the shortest conceivable amount of time, a Planck interval—5.4 x $10^{-44}$ of a second. In one second about 50 million, trillion, trillion, trillion such units have already occurred. Visualizing the instant of time is the same as watching a movie scene by scene. In a motion picture each scene blurs into the next—usually at 24 frames per second—while, watching real time, each moment combines with the next, but at a much, much faster pace, almost $10^{44}$ times per second.

We are thus continually moving toward the future. It never really exists. By the time the so-called future occurs, we remember it as the past. We are always in the present, but all of our knowledge, all that we see, feel, or are aware of, has already happened.

*Question:* You are getting too mystical for me. What are you really driving at?

## Appearing And Disappearing—New And Old Instants

*Answer:* Remember, the fastest possible motion in our universe is the speed of light—$c$ (3 x $10^8$ m/s). Nothing can travel faster, and nothing of mass can reach that velocity. Only energy in the form of an electromagnetic wave is thought to move at that speed. Thus, for you to see anything about you, or distant from you, that electromagnetic wave (light wave) has to travel a certain distance at a set rate. The shortest interval it can attain is a Planck length (1.6 x $10^{-35}$ of a meter), and the shortest time is a Planck moment (5.4 x $10^{-44}$ of a second).

Now we showed before that our universe disappeared at an angle of just greater than 89.99..9$^{59}$ degrees (one Planck length, or moment, less than 13.80 billion light-years). At that point, light would have redshifted to so great an extent (between $10^{32}$ to $10^{33}$ times) that the actual light wave would be longer than the length of the universe (1.3 x $10^{26}$ m) and, thus, our world would be invisible. It would no longer exist as we know it.

I have also stated that the total expanse of the universe must always be just less than $90^0$ on this fourth-dimensional sphere. Therefore, if a new instant is continually occurring, if a new Planck moment keeps appearing, if time keeps on slipping into the future, it has to keep on disappearing from the past.

The universe is the exterior of a rotating fourth-dimensional sphere; we are that third-dimensional surface. We understand the movement into this fourth dimension as time; we cannot visualize a fourth-dimensional direction. But we know that we are always at the present moment. We know that everything about us, everything of which we are composed, is in the past. We think we are in a world of three spatial dimensions, but we actually exist in one of four, the fourth spatial plane thought of as time.

*Question:* You seem to be a little mixed up with space and time. Space is what is about us. Time is a different concept. Of course, since Einstein, we must talk of *spacetime*, but really space, with its three dimensions, and time, must be different, must they not?

## Distance, Time, And The Speed Of Light

*Answer:* Stop for a moment and think—of course I mean this only figuratively, not literally, as you cannot really stop for a moment, but take a deep breath and consider what I have been saying. The universe moves on a moment-to-moment basis, each interval being a Planck instant (5.4 x $10^{-44}$ s). It also moves a Planck length, but into a higher, unknowable, spatial direction. Thus, the universe moves about on a great fourth-dimensional sphere at $c$ (the speed of light). But $c$ is a constant; it is always the same (3 x $10^8$ m/s). Therefore, velocity, a ratio (distance/time—miles/hour, kilometers/second) becomes a constant, $c$; and, since:

$$Distance/time = c, then,$$
$$Distance = (c) \ x \ time.$$

Now, since $c$ is constant, never changing, in reality distance is proportional to time, or we can say they measure the same thing.

*Question:* I guess what you are saying is not really that strange. When we ask visitors en route to us how far away they are right now, they may equally say 40 miles or 1 hour. In this simple case, moving at 40 miles/hour, one can really be separated by 40 miles or 1 hour. I suppose that is what you are trying to show, that, at a constant velocity, distance becomes time.

*Answer:* Yes, that is as good an example as any and is quite easy to understand. People often equate the two when noting they are hours, or minutes, away. In the universe, which moves at *c*, each Planck distance is really a Planck time, and vice-versa. We sense the world as having only three dimensions. It really has a fourth spatial plane, but we cannot visualize it so we instead conceive of it as time.

### What We Visualize—What Is Real

Let me again draw our universe in a very simplified one- and two-dimensional way. Always remember, in our drawing, the one-dimensional straight line is our three-dimensional world, and the two-dimensional circle is our unknowable fourth-dimensional sphere:

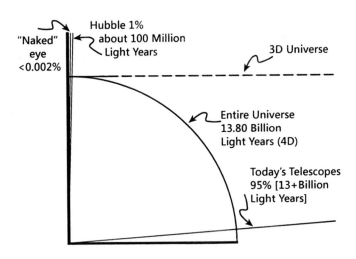

We are drawing a universe of 13.80 billion light-years. Hubble perceived less than 1 percent—only 50 to 100 million light-years. With today's telescopes we see over 13 billion light-years—about 95 percent of the total. With the naked eye, we glimpse very little—roughly 1/50,000 or 0.002 percent of the distance. Remember, we witness only the past. So, unaided, we can grasp around 200,000 years of what has occurred. At that distance/time there is barely any fourth-dimensional curve; thus, we essentially observe a flat three-dimensional universe.

Hubble began to spot the curve but misinterpreted it as velocity increasing with distance. Today we notice it quite distinctly but keep misconstruing it as an ever-greater expansion. But all it really consists of is fourth-dimensional space understood by us as time.

Now, time has a past, a present, and a future, and we are always at the present moment, approaching the future but with knowledge only of the past. If we use the drawing of a straight line to connote our three-dimensional existence, we are always at the tangent to that circle, at the point of connection where the lower-dimensional line touches the higher-dimensional object. We can never move in that direction, as it would be toward an unreachable plane—incomprehensible to us as a spatial dimension.

However, we know that time keeps flowing, slipping into the future. We also know that the smallest unit of time is the Planck interval ($5.4 \times 10^{-44}$ of a second). Thus, the point of connection, the smallest point possible, is also that moment. At that instant, the universe is three-dimensional, moving toward the fourth, the point of connection of a straight line to a circle.

Therefore, in that and every Planck instant the entire universe must exist. The reason we know this is because *we* exist. We have to use that as our most basic concept. Our presence, the universe's reality, has to be a given. If we cannot agree on this, we cannot determine any other thing.

Since the universe must be real, the shortest time frame for its actuality is a Planck interval. In that moment and in every succeeding one, it is fully present. Hence, it must re-establish itself (three-dimensionally) every Planck moment; for, to us, between moments there is nothing—

no time, no space, just fourth-dimensional emptiness. Thus, it continually renews itself; about 50 million, trillion, trillion, trillion times each second.

## 3D Becoming—4D Being

*Question:* Are you saying that we are the edge of a fourth-dimensional curve, and thus, at any Planck instant, three-dimensional? Is that why we see ourselves as we do?

*Answer:* Yes, we are always in the instant of becoming. That interval is the smallest possible time in our universe. For that moment, we have not yet traveled into the next spot on the fourth-dimensional curve. For that brief period, we are just three-dimensional. But in the next instant, we have advanced a Planck length on the way to a higher level. Since this movement is toward the fourth-dimension, and since we cannot conceive of this spatial plane, we instead visualize it as a motion in time.

Because the distance is so small, the moment so short, we do not feel this effect. We are unaware of it happening. All we are aware of is that time keeps passing. All we know is that, as each second goes by, about 50 million, trillion, trillion, trillion instants are constantly becoming. Things continually change; we and everything about us ages. But the universe is just turning in an unknowable spatial direction. It is not aging or growing, it is always the same—measured in time as 13.80 billion years.

## Change—The Only Constant

*Question:* What you say sounds similar to some ancient Greek philosopher who understood change to be the only constant in nature.

*Answer:* Yes, I do not profess to have any original philosophical ideas. I am just trying to give a more reasonable explanation to what is actually observed. Thus, at any Planck interval, the entire universe has to exist. Between Planck instants, nothing as we understand it, or can observe, exists.

# 15

# GRAVITY

G RAVITY, AS IT RECONSTITUTES *the entire universe each Planck instant, must be an "instantaneous" force. Gravity, thus, travels the entire distance of the universe (1.3 x 10²⁶ m) each Planck moment (5.4 x 10⁻⁴⁴ s) at 2.4 x 10⁶⁹ m/s. However, the speed of light (3 x 10⁸ m/s), the fastest possible rate in our world, is much, much slower. The difference is 8 x 10⁶⁰, the same number as Planck lengths making up the universe or Planck moments since the "beginning" of time.*

*Although gravity is instantaneous (immediately re-establishing our three-dimensional world), actual motion (from place to place), by definition, must take time; thus, movement occurs in the fourth dimension (our concept of time). Therefore, the speed of light connotes activity within the fourth dimension, gravity within the third.*

*Another way of conceptualizing this is to visualize our universe as composed of 8 x 10⁶⁰ distinct three-dimensional worlds jittering by, at light speed, over a fourth-dimensional sphere. Each world is reconstituted every Planck instant, with only the very slightest of differences that would appear to occur at the speed of light.*

## Gravity—Instantaneous Force

The force that holds our world together, as described by Newton, is gravity. Therefore, gravity has to re-establish itself, has to remake the universe, every Planck instant. Since the universe encompasses 13.80 billion light-years, if we interpret this as distance, not time, its size is 1.3 x $10^{26}$ m. But the Planck instant is 5.4 x $10^{-44}$ s. Thus, the velocity at which gravity has to travel to reconstruct our world instant to instant is:

$$1.3 \times 10^{26} \text{ m } / 5.4 \times 10^{-44} \text{ s, or}$$
$$2.4 \times 10^{69} \text{ m/s.}$$

However, light speed, the fastest possible speed in the universe, the speed at which only energy can move is much, much slower (3 x $10^8$ m/s). Light speed is less by a factor of 8 x $10^{60}$. When you have a multiple that great—eight trillion, trillion, trillion, trillion, trillion times—it is immediate. Consequently, the speed at which gravity acts is, for all purposes, instantaneous—much, much faster than the universal limiting speed of light.

**Question:** But how can that be possible? Light is how we measure the universe. Its velocity gives meaning to space and time—light-years.

**Answer:** You are correct, but remember, gravity, the force that holds everything in our universe together, must reconstitute our world every Planck instant and, thus, must travel at over $10^{69}$ m/s. Therefore, if one were to suddenly remove the Sun from our solar system, the loss of gravity would be instantaneous (our orbit would be immediately disrupted), but the actual disappearance of its light would take about eight minutes.

**Question:** That is very weird. Are you really quite sure of this concept? It seems to go against all fundamental ideas.

## 8 x $10^{60}$ Separate Universes

**Answer:** Let me clarify it a bit. It has to do with how we understand the third and fourth dimensions. We are always in the present instant, approaching the future. We are always in the third dimension, the edge of the rotating surface of a fourth-dimensional sphere. Each instant we enter into the fourth spatial dimension but think we are traveling in time.

If we were to add up all the moments since the *beginning* of the universe (13.80 billion years), there would be $8 \times 10^{60}$ such intervals. Thus, there are really $8 \times 10^{60}$ separate third-dimensional worlds that make up our fourth-dimensional universe.

*Question:* You are starting to confuse me. What are you trying to say? That in each instant an entirely new world is formed, hence there has to be an instantaneous force that immediately re-establishes everything (at a rate much faster than the speed of light)?

## Instantaneous 3D—Measurable 4D

*Answer:* Yes, that is what I am trying to say. We are the third-dimensional surface of a fourth-dimensional world. We are always at the present instant—in the third dimension—moving toward the fourth. Since each interval is separate, our three-dimensional universe must reconstitute itself moment to moment. We sense this as gravity, and it occurs within that moment, instantaneously.

But we understand the universe as existing in real time, in the past. The past goes back 13.80 billion years. Thus, there are $8 \times 10^{60}$ independent moments, and what is instantaneous (gravity), and sensed by us immediately throughout the universe, when appearing as energy (electromagnetic waves), travels at a much slower rate—the speed of light.

*Question:* I still do not understand. What are you trying to say about third- verses fourth-dimensional velocity?

*Answer:* Each one instant (each $5.4 \times 10^{-44}$ s) reestablishes our existence. The velocity of gravity, therefore, is the velocity of the force that reconstructs our three dimensions. It occurs in that and in every other moment and it is instantaneous.

But our universe really exists as the surface of a fourth-dimensional sphere. We know that as we see redshifting to an ever greater extent the farther out we look. So everything we see and know is found in the fourth dimension. Since we cannot sense this as a spatial direction, we consider it time.

The force we feel moving through this fourth dimension, the force that carries energy, that transmits electromagnetic (light) waves, must journey through time. It travels moment to moment over many third-dimensional instants. Each moment it traverses an entire independent universe. Therefore, although its velocity is 2.4 x $10^{69}$ m/s, it has to go through 8 x $10^{60}$ separate universes; hence, its velocity also is really:

$$2.4 \ x \ 10^{69} \ m/s \ / \ 8 \ x \ 10^{60}, or$$
$$3 \ x \ 10^{8} \ m/s \ (speed \ of \ light).$$

Thus, even if, in third-dimensional terms, the force occurs immediately, in fourth-dimensional terms it has a measurable rate—the speed of light.

## Speed Of Light

Now, the speed of light is a very interesting velocity; nothing that has substance can travel at this rate. It is the limit in our world as it is, in effect, the speed at which our universe moves. Our world is the surface of a fourth-dimensional sphere. If we take any three-dimensional moment, and there are 8 x $10^{60}$ such independent moments, and if we consider that one moment our entire universe, then the speed of light (the velocity of the force that carries energy throughout our universe) would increase by that factor; it would go from 3 x $10^{8}$ to 2.4 x $10^{69}$ m/s.

Thus, gravity travels *within* three-dimensional instants and light *between* three-dimensional instants, over fourth-dimensional space—over time. They have the same velocity, as fast as possible, since they are really the same. But gravity, in three-dimensional terms, is instantaneous, whereas light, in four-dimensional terms, is measurable.

Therefore, no matter how fast something of substance travels, light (being really instantaneous) moves by it at the same constant rate. This strange aspect is the basis of Einstein's special theory of relativity, and it helps us to understand some of the weirdness of quantum theory (*spooky* action at a distance), and also inertia (both of which we will later attempt to discuss).

*Question:* I only vaguely grasp what you are driving at but have to assume that you comprehend it, so let us continue. Throughout the book, you have been talking about the universality of force, how all forces are the same. Why not discuss that now?

# 16

## UNIVERSAL FORCE

P ROTONS REPRESENT INDIVIDUAL UNIVERSES. *The force that holds a proton to-gether, the strong nuclear force (SNF), is $10^{41}$ times as great as gravity, the force that binds our universe. However, our universe is $10^{41}$ times as large as a proton; therefore, both forces are the same.When we gaze at a proton, we are looking at the portal to an individual world but separated by a higher dimension. It is similar to ours, only much smaller; we cannot travel, we cannot visit. However, it is already a part of us, just as we are a part of it.*

*If our universe is the 3D surface of a 4D sphere spinning about at the speed of light (c), gravity can be equated with centripetal force ($mv^2/r$). Thus, m is the mass of our universe, v equals c, and r is the radius of the 4D sphere. Hence, gravity is an inwardly (4D) directed force containing the entire mass of the universe mov-ing at light speed ($mc^2$).Therefore, gravity is the equivalent, or source of all energy in the world ($E = mc^2$).*

*Finally, the electromagnetic force (EMF) is estimated at $10^{39}$ times that of gravity, but like SNF it too is really the same ($mc^2$).The Fine Structure Constant, an unusual and intriguing number ($1/137$, or about $10^{-2}$), is, among other things, the ratio of EMF/SNF. Thus, $10^{39}/10^{41} = 10^{-2}$ and gravity, SNF, and EMF are all one.They all measure the same thing; gravity in 3D, SNF and EMF in 4D.*

## Protons—Individual Universes

*Answer:* Let me try to explain my concept of a *single force*. Gravity is what holds the universe together. It is the attraction that keeps us on the surface of our planet. It is what molds the solar system and maintains it within the Milky Way. Newton elegantly described the concept; Einstein further refined it. Then there are the other forces—electromagnetism, and the strong and weak interactions. The problem with unification is the tremendous differences in their relative strengths.

The strong force is estimated in different ways at $10^{36}$ to $10^{46}$ times as powerful as gravity; most assessments range about $10^{41}$ (or one hundred thousand trillion, trillion, trillion) times as great. Electromagnetism is thought to be approximately $10^{39}$ times, and the weak interaction around $10^{35}$ times the intensity of gravity. Now there is an accepted theory linking the electromagnetic and weak interactions (the electroweak theory) so we can talk of three distinct forces (gravity, the strong, and the electroweak).

Just as gravity holds our universe together, the strong interaction holds atomic nuclei in place. It is what keeps the supposed constituents (quarks) of nucleons (protons and neutrons) in situ and all within a nuclear core. Since we estimate the size of the universe at approximately $10^{26}$ meters, and a proton's diameter at about $10^{-15}$ of a meter, the universe is, therefore, more or less, $10^{41}$ times as large as a proton. If the force holding the universe together were equal to that maintaining a proton, then, since it is diluted over an expanse of $10^{41}$ times, its strength should be diminished by that amount.

*Question:* Wait a minute—are you equating the universe with a proton? How can we be as big as we appear yet as small as an atom's core?

*Answer:* I am saying that, if the attraction—gravity—were compressed $10^{41}$ times, it would be as great as the strong force. Therefore, gravity and the strong interaction should be the same and, yes, each proton thus represents a separate universe. Each proton becomes the portal to a fourth-dimensional sphere replete with a three-dimensional surface world moving in an unknowable direction at the speed of light.

Remember, we reside in four spatial dimensions but think there are only three. We consider the fourth to be time. If we adopt the perspective of an unknown fourth direction, objects our size, but distant, appear to us as we are but much, much smaller and within that of which we are composed.

## Distance Into The Fourth Dimension

Let me give you an example. Assume we are on the top floor of a tall building, looking down at the street below. We see people moving about. They look and act just like you and me; they are simply much smaller. You know that, if we were to leave that building and walk among them, they would be the same as us. The only difference is their *apparent* size; however, as it is based on distance in our normal three-dimensional world, we can move to them and become one of them.

When we gaze toward an atom's center (a proton) we can sense a universe just as ours is only much, much smaller; it is far away but in a direction toward which we cannot travel—the fourth, or inward, dimension. If we could, if we could move in that foreign direction, we would find it just as ours is.

Thus, distant universes appear to us as protons embedded in our world. However, to them we are also but one of many protons set in theirs. It is estimated that there are almost $10^{80}$ protons in total; thus, if each is the doorway to a separate universe, there are at least that many distinct and different destinations.

## Multiple Universes

*Question:* I have heard of theories regarding multiple universes; however, I thought they were based on the expansion of our world due to some kind of continual inflation.

*Answer:* You are talking about multiple universe, or *multiverse* theories. They are based on the notion of a constant inflation. Since everything should continue to inflate, different, innumerable worlds would keep forming, ours being but one of uncountable others. Each, as it was separate from all others, could have distinct properties not found in their brethren; thus, only a few would be hospitable to life forms. Some call this the *anthropic* principle, based on the consideration that there are only a very limited set of physical qualities that are really consistent with the formation of human intelligence. Those would be rather unusual; therefore, according to these theories life, as we understand it, would be quite rare indeed.

However, the whole concept of inflation makes no sense. There was no Big Bang, there is no expansion so inflation is not required; but even if there had been, it is, essentially, infinitely impossible due to ever-increasing entropy. Therefore, if each proton (replete with the strong force) is really a universe (contained by the force of gravity), all worlds would be similar in their basic characteristics. Thus, if ours supports intelligent life, all others could too. However, communication between separate ones is probably impossible, as each is a distinct fourth-dimensional black hole with a three-dimensional event horizon from which, by definition, nothing, including light, can escape.

**Question:** So you are equating all forces. Gravity and the strong force may be the same, but what about the electromagnetic–weak interaction? You were showing some similarities between the velocity of gravity and that of electromagnetic waves, as gravity is in the third dimension (within the instant) and electromagnetic or light waves are in the fourth dimension (between those instants) within time.

But electromagnetic waves carry energy. They interact with the electrons at an atom's periphery—photons are emitted. They can be blocked by opaque barriers. Gravity does not appear to do these things. It certainly cannot be stopped by any obstacle. It just is; it is what mass does.

## Gravity And Centripetal Force

**Answer:** Let me try to more fully explain myself. Newton stated that all matter attracted all other matter instantaneously over any distance by something he called gravity. He did not try to explain what it was comprised of, just that it existed. Einstein further showed that gravity was really a bend in the fabric of spacetime, and that its attractive force depended upon this distortion.

Let me give you an example; it will help to explain my concept. If we take a ball attached to a string and spin it about in a circle, the force felt on the ball is the pull toward one's arm by the string. The ball can be traveling at any velocity, but the fact that it moves in a circle means there is acceleration (change in direction) at a constant rate. Acceleration connotes force (force = mass x acceleration), and the force is inwardly directed toward one's arm holding the string. This is called *centripetal* (or central pulling) force.

Now the formula for this is easily derived, and can be found in most science textbooks. Since you did attend college and took basic physics I know you at least encountered it at some time. The formula is:

$$F = mv^2/r.$$

Force is equal to the mass of an object times its velocity squared, divided by the radius of the circle.

Let us take the formula and place it in the perspective of our world. We are now the ball spinning about the imaginary string. The mass of the object is the mass of our universe. The velocity that we spin about is the speed of light—$c$. Thus the force exerted by this spin is:

$$F = mc^2/r.$$

Now, the radius of the spin is in an unknown or fourth direction; thus, to us it is only imaginary. If we use a mathematical term for an imaginary number, such as the square root of minus 1, we use the term ($i$). Therefore, the force felt in our universe is:

$$F = mc^2/(i).$$

Or we can say:

$$(i)F = mc^2.$$

Since ($i$) is an imaginary direction, it is within all that exists. Thus, the force is *inward*, no matter in which direction one looks, and is equal to:

$$mc^2.$$

Now, $mc^2$ is equal to energy ($E = mc^2$), and since we are describing the mass of the entire universe, we are discussing its total energy pulled toward this inward or imaginary direction.

Therefore, gravity is an attraction at every point in the universe, proportional to the mass at that location. It is energy directed toward the fourth direction, inwardly, and, thus, establishes our world. Electromagnetic force constitutes the same energy ($mc^2$) but diffused throughout the universe, interacting with other individual worlds. It travels through the fourth dimension, through time.

Gravity remakes the universe instant to instant; electromagnetic force is how it is perceived. Electromagnetism allows us to understand how each instant interacts with the next, how the universe flows toward the fourth dimension. It allows for time. But the forces—the inwardly directed energies—are the same; they both equal $mc^2$.

**Question:** But if both are similar why does electromagnetism appear $10^{39}$ times as great?

## Fine Structure Constant

**Answer:** There is a ratio in physics known as the *fine structure constant*—$1/137$. It can be derived in several ways, but one of its meanings is the relationship between electromagnetism and the strong force. Now, remember, we are stating that gravity and the strong interaction are really the same; gravity is merely diluted by the size of our universe in comparison to the size of a proton, or about $10^{41}$ times.

Therefore, if the strong force is 137 (essentially $10^2$) times as great as electromagnetism, and also $10^{41}$ times as great as gravity, it stands to reason that the electromagnetic force is $10^{39}$ times that of gravity ($10^{41}/10^2 = 10^{39}$). Thus, all the forces represent the same thing (they all measure centripetal attraction, which allows for rotation of three-dimensional surfaces on fourth-dimensional spheres—$mc^2$).

However, gravity measures the force in three dimensions, instantaneously, whereas the electromagnetic force is determined in four, over time. The vast discrepancies in strength, then, are due to the different dimensions these forces describe.

## SECTION IV

# Black Holes, Holograms, And Entanglement

T HE FOLLOWING CHAPTERS DISCUSS some concepts concerning black holes as 4D structures. Each proton represents a portal to another world. Our universe is $10^{41}$ times as large as a proton; thus, if our world is understood as the surface of a great 4D sphere, that entity's radius could contain about $10^{41}$ protons, its surface roughly $10^{82}$ and its volume around $10^{123}$. This is the basis of the very intriguing "large number hypothesis."

Another way of visualizing the entire universe is as a hologram; its surface—our 3D world—then has all the data or knowledge found within that entity. It has also been shown that a quarter of the surface of a black hole encompasses all the information contained within that body, and our world is always just that exact one-quarter.

Virtual particles always come in pairs—the particle, and its antiparticle. This is because we are really observing the same 4D object from its "outside" and "inside"—we see it both normally and inside-out. Antiparticles are also quite rare, as it is only in unusual circumstances that we can observe something from a 4D perspective.

Finally, as all protons represent other, entire worlds, we become but a tiny proton within innumerable distant universes. Thus, every particle contains every other particle—we are all entangled.

# 17

# BLACK HOLES AND LARGE NUMBERS

A PROTON PRESENTS AS A MINI-UNIVERSE; *therefore, it is the equivalent of a black hole with a 4D spherical center and 3D skin or event horizon. A much greater black hole occurs when the gravitational force of a star overcomes all resistance and collapses that object's contents to a point where nothing, including light, can escape. Its center becomes the home of innumerable individual protons' centers, and its surface one great third-dimensional event horizon. It, like the proton, is an individual universe just as are we, all intertwined together.*

*The large numbers $10^{40}$, $10^{80}$, and $10^{120}$ are considered by some to have a profound significance; they appear as ratios in many important, basic concepts. The universe is $10^{41}$ times as large as a proton; therefore, if we are the 3D surface of a great 4D sphere whose radius is about $10^{26}$ m, $10^{41}$ protons edge to edge would equal that radius. Thus, the surface—our universe—would contain around $10^{82}$ protons ($10^{41}$ squared), and its volume would hold approximately $10^{123}$ protons ($10^{41}$ cubed).*

*Since there are thought to be about $10^{80}$ protons in the visible universe, that equates with the surface of this immense sphere; and since the ratio of theoretical vacuum energy to measured energy is about $10^{120}$, if each proton is a universe, with all the energy of ours, then there should be, more or less, $10^{123}$ potential protons in the great vacuum of empty space, equating with this tremendous theoretical energy.*

## Black Holes

*Question:* You were discussing protons as portals to other worlds distant from us in another dimension. You based this on the similarity of the strong force and gravity, claiming their strengths would be equal if they were both measured in the same-sized entities. What about black holes? Aren't they thought to be at the center of galaxies? Aren't they also, therefore, fourth-dimensional concepts? Could they too be universes just like ours?

*Answer:* Black holes are believed to form when gravity overcomes all other forces. Nothing can escape from these objects, not even light; thus, they are described as black. They are theorized to be at the center of all galaxies. Some even think that they may be at the core of stars. Since there are estimated to be 300 billion trillion stars, there are potentially many, many black holes.

They are found when a star's energy sources diminish and it can no longer resist gravity's attraction. An immense explosion—a supernova—occurs, blowing off the surface, leaving a very dense core. If the star, prior to this explosion, was greater than ten times the mass of our Sun, this material is so compressed that only a black hole remains—a fourth-dimensional object replete with third-dimensional surface (event horizon).

This black hole would be analogous to the creation of the ninety-plus elements that are formed from hydrogen. In stars, hydrogen is continually fused into helium, which can then combine to more complex matter (when stars explode, in supernovas, all the elements are finally forged).

When a black hole is produced, a similar process occurs, except that all the higher-dimensional centers of the nuclei fuse into one immense bulk. This becomes the equivalent of a great atomic nucleus, a black hole with event horizon, the equal of many, many protons. Thus, one could describe these black holes as giant worlds composed of a great number of smaller universes or protons.

*Question:* So, if I understand what you are saying, a proton represents a universe, and a black hole, since it is an aggregation of many, many protons, would be equivalent. It would be much, much larger than a single proton but would essentially represent the same thing.

*Answer:* Yes, that is what I am trying to show. Protons are portals to other worlds. Conglomerations of protons' centers—large black holes—would also be similar. The force that holds protons together, that holds nuclei together, that holds black holes together, is the same force that holds our universe together—gravity.

## Protons—Mini Black Holes

Let's return to the concept of the proton as a doorway to a universe similar to ours but distant into the next dimension. The proton, thus, appears to us as a black hole but very much smaller than those found at the center of galaxies. At its core is a higher-dimensional construct, a fourth-dimensional sphere. This structure would not exist in our three-dimensional world; it resides in the higher plane. However, since it is a fourth-dimensional creation, a sphere, its surface is what contributes to our understandable universe. The exterior of this object is three-dimensional, just as any normal three-dimensional globe has a two-dimensional skin. The cover is always one dimension less than the object it is enclosing.

Now, the proton's diameter is approximately $10^{-15}$ m. The next tangible substance, our electron cloud, goes out to about $10^{-10}$ m, or a distance 100,000 times as great as the proton core. Subsequent concentric shells are ever-more tenuous; however, they continue to the very edges of existence. Each proton is, therefore, the portal to a universe that encompasses all others. In fourth-dimensional terms, we are but a proton embedded in every other world and, as already noted, about $10^{80}$ protons (or worlds) exist.

## Large-Number Hypothesis

Paul Dirac, an influential mathematician who was instrumental, in the 1920s and '30s, in the formulation of much of quantum theory, stated that there was something fundamental to the large numbers $10^{40}$ and $10^{80}$. He suggested that it was not mere coincidence, not simply numerology, that these integers were the origins of several important scientific ratios. He felt that there had to be a deeper significance to them.

*Question:* Well, you have been discussing similar large numbers. You have noted the ratios of the strong force to gravity, and the universe to the proton as both $10^{41}$. You have given the number of protons in the universe as $10^{80}$. Do you see any relation to these proportions and Dirac's claim of significance?

*Answer:* Let me explain my concepts. Remember, we are stating that the tangible proton is the event horizon or three-dimensional surface of a fourth-dimensional sphere. It is about $10^{-15}$ meters. Thus, $10^{41}$ protons, edge to edge, would equal the diameter of a much, much larger entity (approximately $10^{26}$ m) whose surface is our universe. The exterior of that great sphere—our world—would then contain around $10^{82}$ protons ($10^{41}$ squared). This is roughly the same number as the total protons estimated to make up our world.

Finally the total volume of that sphere would be in the order of $10^{41}$ cubed, or about $10^{123}$. This extraordinarily large number is similar to quantum mechanics' computation of the enormous theoretical vacuum energy of the universe. (That estimate, of the entirety of the potential energy of the universe, is calculated at, more or less, $10^{120}$ times what is routinely noted—the total of all energy and matter, both observable and dark, present in our world. This tremendous discrepancy of $10^{120}$ times reality has been considered by some to be the *worst* theoretical prediction in the history of physics).

*Question:* So you are saying that the large number hypothesis of Dirac ($10^{40}$ and $10^{80}$), and the $10^{120}$ value given by vacuum energy, make sense as the radius, surface, and volume of a giant fourth-dimensional sphere. That sphere contains our universe, its surface is equal to every existing, perceptible proton, and its volume equals all latent or potential protons if they were to entirely fill our world.

## Enormity Of The Vacuum

*Answer:* That is exactly what I am saying. Look, each proton represents an entire universe similar to ours. Each proton establishes the three dimensions of our world; initially with an event horizon ($10^{-15}$ m), then concentric orbs (starting with electrons—$10^{-10}$ m), and finally extending to the very edges of space ($10^{26}$ m). Remember, our universe is but a three-dimensional skin. Because of that, it is made up of particles of ever-increasing size (spheres within spheres) to its outermost reaches; however, this is hard to visualize, since we are part of it.

Just as any three-dimensional object has a two-dimensional exterior, a four-dimensional entity would have a three-dimensional surface. However, we are that surface, and everything we know or understand is part of it. Thus, if the diameter of the universe contains about $10^{41}$ protons, then the surface, in all its nooks and crannies, holds approximately $10^{82}$ protons. That is why there are estimated to be around $10^{80}$ protons in the known universe (about the same as $10^{82}$). Protons (with their associated concentric ever expanding spheres) are what make up our world.

The total theoretical vacuum energy is around $10^{120}$ times as great as the actual *dark* energy and matter. But remember, this is about 20 times as much as the actual energy of all the *observable* particles. Thus $10^{120}$ times as great as dark energy–matter is closer to $10^{122}$ times that of all the visible energy in the universe (a number very similar to our estimated volume of the universe—$10^{123}$). Thus, the theoretical energy of the universe—quantum vacuum energy—is the equivalent of $10^{123}$ separate worlds packed into one.

## All As One, One As All

*Question:* I see where you are leading. Dirac was correct; $10^{40}$, $10^{80}$, and $10^{120}$ are fundamental ratios; they are not just coincidental numbers. They are due to the fact that the universe appears as a proton and each proton represents another world. However, one has to look into the next dimension to see that.

Therefore, the force of gravity is really the strong force, only diluted

$10^{41}$ times. The number of tangible protons is what makes up the three dimensions of our world, the skin so to speak, and, thus, there are $10^{82}$ (approximately $10^{80}$) such entities. Finally, the extensive volume is filled with potential or virtual protons, each with the energy of the entirety. Hence, the latent energy (quantum vacuum energy) is about $10^{123}$ times as great as actually measured.

*Answer:* You really do understand. Dirac was right; the large-number hypothesis is not numerology, not just a coincidence. It is based on a higher dimension in which each entity is a universe (replete with a three-dimensional exterior or event horizon) but represented in every other world as a proton (with attached electron clouds, then evanescent shells extending to that world's very edges).

Remember, we are the three-dimensional surface rotating about a fourth-dimensional sphere at the speed of light traveling one Planck length each Planck instant. But every moment, an entirely new universe is created, formed by the instantaneous action of gravity. Thus, our world is but one Planck length in size—fourth-dimensionally—and yet 8 x $10^{60}$ times as large third-dimensionally.

# 18

## HOLOGRAPHIC UNIVERSE
## AND INFORMATION

H OLOGRAMS ARE TWO-DIMENSIONAL surface representations containing all the information of a depicted three-dimensional item, and each individual part of that hologram has the complete knowledge of the entire object. Some scientists think that our universe is such a hologram.

What we really consist of is a three-dimensional hologram representing a fourth-dimensional sphere. All the information in that sphere is found on the surface—our universe. There supposedly can be only $10^{122}$ individual bits of information making up our entire universe, a number approximately equal to the total theoretical vacuum energy ($10^{120}$) or to the total potential protons in empty space ($10^{123}$). Thus each proton, potential and real, has all the information and all the energy of the entire universe; all are intertwined or entangled.

Also, it is theorized that all the information of a black hole is on a quarter of its surface—its event horizon. We are always that one-quarter of the surface of a fourth-dimensional sphere, never more, never less. Thus, it very intriguing to note that all the possible information of the entire universe must be within us.

### Holograms

*Question:* I remember having read somewhere that there are scientists who feel the universe can be best explained as a cosmic hologram.

Now, I have seen holograms in movies and even in Disney World. They are very weird, as full three-dimensionality is obtained from a two-dimensional creation.

*Answer:* There is an intriguing theory that we are the two-dimensional holographic projection of a three-dimensional universe. In a hologram all the information of the three-dimensional entity is found on the two-dimensional surface. If that interface is sectioned into smaller pieces, no matter how minute, each still has all the knowledge of the original object.

These theorists depicting the universe as a hologram are onto something; they just do not quite grasp it. We are the exterior of a fourth-dimensional sphere; thus, we would be a three-dimensional hologram of a higher reality if they are correct.

All the information of the fourth dimension is found on its three-dimensional surface (the event horizon of a black hole). Thus, since each tangible proton, each universe, is but that event horizon (the exterior of a fourth-dimensional sphere), each proton universe has all the data of its entire enigmatic globe. Also, if holographic theory is correct, each proton has all the information of the entire universe consisting of all other protons—our entire fourth-dimensional world—as each surface part contains all the knowledge of the whole.

## Information

*Question:* You were just saying that all the information of any mass can be found on its surface. What do you really mean by "information"?

*Answer:* Information is not what you may think it is. It is not an answer to some question or some insight into a problem. It is the physical properties of some object measurable through electromagnetic radiation by another entity. Thus, all the physical characteristics of a fourth-dimensional sphere should be measurable on the surface of that orb—on its event horizon.

If the universe is such an object, its radius would be 13.80 billion light-years, or $8 \times 10^{60}$ Planck lengths, and its exterior would be approximately that squared or could contain up to about $10^{122}$ potential bits of information. But this surface is also our knowable world, and if our world were a hologram, each smallest bit would have all the information of the whole.

**Question:** Wait a minute. Aren't you getting a bit sloppy in your estimates? Before, when discussing tangents and redshifts, you gave the radius of the universe as 8.785, not 13.80, billion light-years.

**Answer:** You are right—I am becoming a little loose with my numbers. If we wish to be more consistent, we can use 8.785 billion light-years, or $5.1 \times 10^{60}$ Planck lengths; but it really makes no difference. We are not looking for exact measurements; we are simply rounding off numbers so that a concept can be understood. The shorter length ($5.1 \times 10^{60}$), when squared, still gives about $10^{122}$ units.

More importantly, this is about the same as the total number of protons—real and potential—that make up our entire universe, or the theoretical energy of the vacuum. Thus, each proton in essence would have all the information of the whole, each as a separate universe, but each with all other worlds contained within.

I also find very intriguing what theorists who explore the properties of black holes have shown—that all the data (the entire description of all the particles) of a black hole are fully present on only one-quarter of its surface. This is very interesting, as we have always described our universe as that same one-quarter of the exterior of a fourth-dimensional sphere (or black hole).

Remember our original diagram, with the concept of a three-dimensional universe drawn as a one-dimensional straight line, and the real four-dimensional universe depicted as a circle. We always make up almost exactly one-quarter of that circle. Beyond $90^{0}$ there cannot be anything—it is infinitely far away. At less than $0^{0}$, nothing can exist—it has not yet occurred. Thus, our universe is always just less than a quarter of the surface; it can never be larger or smaller. However, as it is our entire world, it has knowledge—information—about everything that exists.

Let me draw it now, showing our essentially unknowable fourth-dimensional spherical universe as a three-dimensional globe. The two-dimensional surface becomes our three-dimensional world, and we see that it covers but one-quarter of that globe.

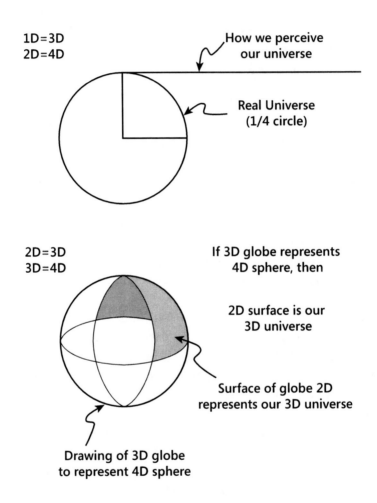

1D=3D
2D=4D

How we perceive our universe

Real Universe
(1/4 circle)

2D=3D
3D=4D

If 3D globe represents 4D sphere, then

2D surface is our 3D universe

Surface of globe 2D represents our 3D universe

Drawing of 3D globe to represent 4D sphere

What took 1/4 of circle
Now takes 1/4 of surface

**Question:** So you are saying that our three-dimensional universe, everything that we know of, everything that exists, is but the exterior of a higher-dimensional orb. Since that sphere is a true black hole, *we are* that covering, the event horizon. All the information, all the knowledge of all the particles that make up that black hole, is found on that surface, on that event horizon. Thus, all the knowledge or information is available to us and it comprises our world.

**Answer:** That is what I am saying, and since we are the surface of some vast black hole, and since we can only embrace one-quarter of that external facade, it is very fitting, indeed, that comprehending just one-quarter is all that is needed to know the entirety.

# 19

## VIRTUAL PARTICLES
## AND ENTANGLEMENT

THE UNIVERSE'S "THEORETICAL" VACUUM ENERGY is believed to be about $10^{120}$ times as great as its "real" energy. This is because the entire energy of the universe resides in each proton, and there are approximately $10^{123}$ protons, real and latent, in the universe. Most are dormant protons existing in the great vastness of empty space—the fourth dimension.

When one "pops" into our 3D world, we first sense its outside; it presents as a particle. As it passes through the skin that makes up our 3D world, we then visualize its inside; we call this an antiparticle—the exact inverse (the inside-out) of the particle. After exiting, the surface re-closes with tremendous energy equal to $mc^2$—the centripetal force of gravity. Therefore, virtual entities come and go, always in pairs, disappearing with a "bang."

Since an antiparticle is really the other side of a particle observed from a 4D perspective, there are very few of them in the real world. They are only found traversing our 3D skin or when a high-energy collision dislodges a previously stable one, pushing it toward the fourth dimension. Each and every proton represents, and has all the energy and information of, an entire universe; they are all interconnected, they are all entangled.

## Virtual Particles And Antiparticles

*Question:* When you were discussing the large-number hypothesis, you stated that, because the universe is filled with dormant entities (unseen, as they reside in the higher dimension), its total (or theoretical) vacuum energy was much greater than what appears real or measurable. What happens, then, if one of these latent objects passes into our three-dimensional world?

*Answer:* According to quantum theory, the vacuum is teeming with virtual particles that suddenly form, then just as quickly disappear. That is why there is supposedly so much vacuum energy—over $10^{120}$ times as much as is present in the visible or tangible world. In this great vastness, particles are always popping into momentary existence, then abruptly vanishing. In fact, whenever one appears, it does so with its antiparticle; they then annihilate each other with tremendous force.

To explain this phenomenon, let us again represent the fourth-dimensional universe as a three-dimensional globe; our world, then, becomes the two-dimensional facade of this orb. The particle that comes into existence can be viewed as puncturing a surface from the *inside* passing through and exiting on the *outside*. Since we have drawn a particle, for the sake of understanding, as three-dimensional, it first touches our inner surface with the outer aspect of its cover.

But remember, since we are using the analogy of a three-dimensional globe, its two-dimensional exterior is our world. Now, inhabitants of a two-dimensional surface have no concept of inside or outside; there is no height or depth, no top or bottom to that plane. Thus, if we transpose this to our real world (a four-dimensional sphere), the place from which a virtual object comes does not exist. It essentially appears out of *nothing*—the next dimension.

Since it touches our world with its outside, we immediately perceive it to be a real particle, a proton or other similar entity, for, once it enters our space, it assumes three-dimensional characteristics. It passes right through this surface and, as it exits, we glimpse its inside. Again, remember, inside and outside have no meaning to us, their meaning resides only in the fourth or higher dimension. Also, because each entity is rotating (as are we) at $c$ through the fourth dimension, we are constantly visualizing their different aspects or phases as they pass through. This constant change in shape and size (to be discussed later) becomes the basis of both wave–particle duality, and spin $1/2$ (the reason for $720^0$, or two full turns to complete one rotation).

Therefore, objects as they exit are exactly like those that entered, but we are seeing them (they project to our surface) from the inside out. Thus, what we consider positive (a proton) now becomes negative, and what was negative (an electron) now turns into its inverse (or positive). Hence, we see the particle on entrance and the opposite or antiparticle on exit; it is, basically, the same particle, but inside out. That is why every virtual particle is seen with its exact replica antiparticle. Let me try to draw this.

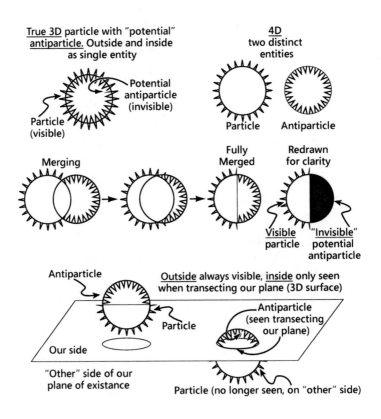

Now, once through the fabric (the facade of our sphere), the hole formed closes with a burst of energy. The energy is proportional to the mass of the particle formed, or $mc^2$. Remember, gravitational force, the force that holds our universe together and leads to the counterforce of inertia and substance, is just energy directed inwardly (toward the fourth dimension) and is equal to $mc^2$.

Therefore, virtual particles come and go, always in pairs, with a particle and its antiparticle and always with a burst of energy. What is really happening is that a surface (our world, our comprehensible three dimensions) is rent by a higher-dimensional object passing through and establishing for the briefest moment a particle and its antiparticle, and then closing with a bang. Let me make an attempt to demonstrate this.

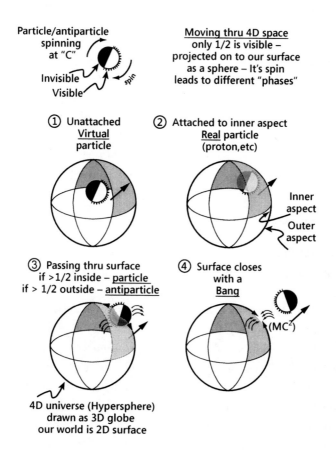

Particle/antiparticle
spinning
at "C"
Invisible
Visible
*spin*

Moving thru 4D space
only 1/2 is visible –
projected on to our surface
as a sphere – It's spin
leads to different "phases"

① Unattached
Virtual
particle

② Attached to inner aspect
Real particle
(proton,etc)

Inner
aspect
Outer
aspect

③ Passing thru surface
if >1/2 inside – particle
if > 1/2 outside – antiparticle

④ Surface closes
with a
Bang

$(MC^2)$

4D universe (Hypersphere)
drawn as 3D globe
our world is 2D surface

*Question:* You know, with your explanation of virtual entities, you try to show why there is always an exact opposite replica or antiparticle. But in the real world there are many, many particles but very few anti-particles. Why do you think that would be?

*Answer:* Remember, we are the three-dimensional surface of a fourth-dimensional sphere. We are composed of protons with surrounding electrons, representatives of other such orbs (hyperspheres) that have stabilized upon (or within) that surface. Thus, only the outside of each entity is present to make up this facade. These are usually characterized in our universe as hydrogen atoms (each with a proton and electron). It is only when such an object (or part of one) travels through this covering that the inside or antiparticle is seen. The act of transecting the exterior (our three-dimensional world) leads to viewing the inside of this fourth-dimensional sphere, the inside-out version of an entity, the antiparticle. Therefore, in the real universe stable particles are essentially all that one finds. Only momentarily traversing virtual particles, or very-high-energy collisions (that knock previously established entities off their perches, the three-dimensional skin we call our universe), will give rise to antiparticles. Thus, they are rarely seen. They are not normal denizens of our world but the fleeting innards of fourth-dimensional spheres.

## Entanglement

*Question:* Let me attempt to summarize some of your thoughts. We are the facade of a higher-dimensional black hole, its event horizon. All the information that makes up that entity is found on one-fourth of that surface. This is our universe, always one-quarter of the whole.

In a sense, however, we can also be viewed as a hologram, and, thus, all the information of the entire fourth-dimensional entity would be found in any part of the three-dimensional covering. Since the smallest part possible has a diameter of one Planck length, and our universe has a radius in the range of $8 \times 10^{60}$ such lengths, its exterior, our known world, would have about $10^{122}$ such parts or bits (that number squared). Each would have within it all the information of the whole.

But there are also $10^{123}$ possible protons in the huge vastness of our universe; $10^{82}$ make up our real three dimensions, but the great, great majority, $10^{41}$ or 100 thousand trillion, trillion, trillion times as many, exist in a virtual state and when found arrive with their antiproton. So each proton, each portal to another world, theoretically has all the information of the entire universe—the entire universe is entangled. Each proton, both real and virtual, represents a universe containing all other protons—all other universes.

*Answer:* You really do understand, and even expand on what I have been saying. The whole quantum mechanical concept of entanglement—all the knowledge of all the parts of the universe in each and every part—is due to the fourth-dimensional black hole center and subsequent third-dimensional surface. If viewed as a hologram, we, as three-dimensional beings, embody fourth-dimensional abstractions. But all the possible information in that mysterious higher dimension is present in our world. Let me attempt to draw it.

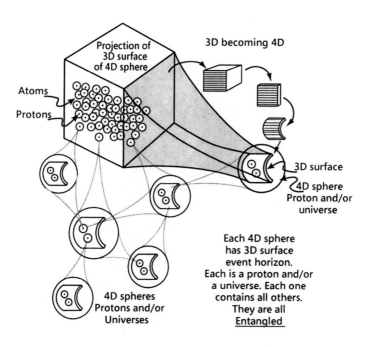

*Question:* So the universe as you understand it is the surface of a black hole. It is three-dimensional and appears flat but is really bent toward the higher plane. It rotates at the speed of light, always in that unknowable direction—it moves toward time. It is placed among other similar rotating spheres—hyperspheres—and each and every one denotes a separate but interconnected universe. We, and they, are all entangled.

## God And The Fourth Dimension

*Answer:* That is a pretty good explanation of how I visualize the universe. Since, as you have noted, we conceive of a fourth spatial direction as time, our past, present, and future would be visible to a higher-dimensional being. From the vantage point of the fourth dimension, the universe we call ours, the three-dimensional skin moving about in an inexplicable direction, our spinning black hole, would be seen by that being as a sphere. Thus, what has occurred, what is occurring, and what will occur would be present to that being at any instant. That being would be omniscient, all-knowing as to our history and future.

*Question:* Are you equating such a being with God? Are you stating that God would be a being in a higher dimension, observing us, knowing what had already transpired and what was to become?

*Answer:* I never would state that God is simply a being, either in our or the next dimension. I feel that God is not only omniscient, all-knowing, but also omnipresent and omnipotent. God is everywhere and the cause of all things. Thus, you cannot place God *within* the fourth dimension; God *is* the fourth dimension, the cause of our actions, the essence of our existence.

The fourth dimension gives rise to force; it engulfs and shapes our world. Before we describe it further, I wish to dwell on a few aspects of the micro or quantum world and electromagnetism. We have already alluded to some of this when discussing virtual particles and entanglement. I want to try to explain several perplexing concepts as best I can. Then I will try to draw my ideas together.

# SECTION V

# *Quantum Weirdness*

QUANTUM THEORY IS DISCUSSED *from a 4D perspective. The two main concepts that make the micro (quantum) world different than the macro (real) world are instantaneous, or "spooky," action at a distance; and wave-particle duality.*

*When looked at from a 4D viewpoint, instantaneous action occurs within one "universe," represented in our world as a proton. Therefore, it is similar to gravity presenting with an immediate transfer of information. Wave-particle duality is how a 4D sphere is visualized in our 3D world. We sense only its reflection, and, as it spins through the higher dimension, different aspects of that shadow appear (over a very short time-frame) both particulate and wavelike.*

*Motion is shown to be illusory as 3D objects, shadows of 4D entities, are frozen or immobile during each instant. However, as the universe (as already noted) moves through the 4D, through time, changes occur; things realign, frame to frame. Therefore, from a 4D perspective the past, present, and future always exist; the future has already happened.*

# 20

# INSTANTANEOUS "SPOOKY" ACTION

QUANTUM THEORY IS DIFFICULT to understand, as it contains concepts that are not apparent in the macroscopic, real world—instantaneous, "spooky" action; and wave-particle duality. In the quantum world, information travels instantaneously over any distance; in the real world, it is limited to light speed. In the quantum world, all substance has a dual wave-particle existence; in the real world, objects can be wavelike or particulate, but not both at the same time.

Since each proton represents an entire universe—a black hole surrounded by a 3D event horizon or skin—information is transmitted over that entire 3D surface instantaneously, just as gravity is in our universe. This 3D surface extends out to the very edges of our universe and encompasses all that exists. All protons have the entire information of our world; all protons contain all other protons within them. Thus, information "within" a proton is transmitted instantaneously to any spot in the universe.

However, when a signal is "between" protons, between individual universes, that information must travel in the fourth dimension through time; thus, that information cannot travel faster than the speed of light. Therefore, spooky action is nothing more than information transmitted within the third dimension, whereas the real world travels in the fourth.

## Quantum Theory—Believing But Not Seeing

Quantum theory arose in the early 1900s on work done by Max Planck and was fully established as quantum mechanics (*mechanics*, in physics, is the study of motion) by the late 1920s. As a theory, it answers many questions about the submicroscopic world of atoms and their constituent parts. But in answering these questions, it makes assertions that are very hard to understand in real-world terms.

The two basic concepts are, first, that all atomic components (protons or electrons, to name just a few) can be described both as waves and/or particles. How one measures these elements then determines their actual characteristics. Second, it allows for instantaneous or, as Einstein stated, *spooky* action at a distance.

These two characteristics of the quantum world are not supposedly present in the real world. Actual things exist and have substance; they do not change due to measurement. If they are solid—particulate—they remain so independent of examination. Discernible cause-and-effect relationships can be perceived. Things have definite characteristics; they are not indeterminate, awaiting some specific measurement to discover their true identities.

Also, in the real world nothing is supposed to travel faster than the speed of light. Thus, in the real world there can be no instantaneous or spooky action at a distance.

**Question:** So quantum theory explains how things actually work in the world of the very small but is unsatisfying to many, as it makes no sense in how we see the real, macroscopic world.

**Answer:** Yes, that is the essence of it. Einstein could never come to terms with quantum theory even though some of his earlier writings formed its basis. His famous admonition *God does not play dice* means things are not haphazard; there must be understandable cause-and-effect relationships.

Yet quantum theory has been proven correct time and again. Experiments have shown that submicroscopic material can, at the same time, be either wavelike or particulate (depending on the measurement criteria), and that information can be transferred, over any distance, instantaneously. Thus, quantum theory is correct; it merely does not make sense in our everyday realm.

*Question:* Well, you have been discussing the fourth or higher dimension, and the way size in our world is dependent on distance into that unknowable bulk; can these concepts perhaps throw some light on the quantum world?

## Size Matters

*Answer:* I am glad you bring this up; size matters. Quantum concepts can be visualized, and made sense of, if size in the third and fourth dimensions can be understood.

When dealing with atoms and subatomic particles, we are going to simplify our discussion by only using hydrogen atoms. These are the simplest to understand, since they have but one proton and one electron. They also make up about 75 percent of all material in our universe and are the basis of all other entities—their fusion, in the cauldron of the stars, to more complex elements is the cause of all other tangible things.

In hydrogen, the nucleus has a single proton. We have already described a proton as a portal to a fourth-dimensional universe, similar to ours, only much smaller as it is far away in that unknowable bulk. It presents to us as a black hole. We have also already stated that such an entity, when fully visualized in three dimensions, appears as spheres within spheres to the edges of our world, and that these concentric shells do not have to be of palpable matter but can be of energy, because energy and matter are the same.

## Protons And Their Universes

Now, since each proton represents a universe, the concentric orbs emanating to our outer reaches are what constitute our world. Our universe is a conglomeration of many, many other worlds—all interconnected. Each presents as an atom (proton center, electron cloud) but extends to the very edges of existence.

Remember, we are only the three-dimensional exterior of a mysterious sphere rotating at $c$ toward time. Therefore, from a higher-dimensional perspective, our entire universe (over $10^{26}$ meters) has the smallest possible thickness; it is the cover of an unknowable globe. Yet this thin lower-dimensional skin is, to us, everything.

All fourth-dimensional spheres, all centers of protons, are displayed on, and form, this surface. There are approximately $10^{80}$ such centers, and, since we visualize them in three dimensions, each becomes a black hole with event horizon and tenuous shells to the edges of what we consider our world. The boundary to us is always that ineffable nothingness of a higher plane.

Each, then, is a separate universe, but we are all part of each other; we are all entangled. Their electron clouds and elusive spheres of matter–energy go out to our borders, to the emptiness of the fourth dimension, just as we go out to theirs.

As gravity instantaneously re-establishes our world, a signal reorienting any atom is also recorded immediately (to the outermost reaches of our world). Therefore, when we discuss instantaneous (or spooky) action at a distance, we are simply noting gravitational force, but with respect to an atom.

*Question:* I am beginning to understand what you are saying. The fourth dimension is a distance that to us appears imaginary, yet it is real. It can only be sensed as time. It causes gravitational force due to the rotation of third-dimensional surfaces on fourth-dimensional spheres. There are many, many individual spheres (estimate, $10^{80}$) spread over the surface of our fourth-dimensional orb, and each represents a universe unto itself, yet each is part of every other world.

When we visualize a foreign universe, it appears as the nucleus of a hydrogen atom (a higher-dimensional sphere with its surrounding lower-dimensional surface, a proton). That object, with its multiple concentric spheres (electron clouds and matter–energy orbs), extends to the very edges of existence, with instantaneous force toward its center.

All realms (our three-dimensional world and its multiple brethren) are composed of the entirety of these concentric shells. They are all entwined. The force of establishment is transmitted so very, very rapidly that action at a distance, gravity or spooky quantum action (entanglement) appears, in essence, to be instantaneous.

## Multiple Interconnected Universes

*Answer:* Yes, you do see what I am trying to show. Instantaneous or spooky action at a distance occurs within one three-dimensional universe; it remakes it moment to moment. But all are interconnected—entangled—since each world is a combination of many, many such entities. Each becomes a proton when visualized in three-dimensional terms, and each is but a part of every other.

Remember, gravity causes the immediate reformation of our universe every Planck moment; it is an instantaneous force. It acts within our three dimensions. Thus, every moment, these concentric spheres of matter–energy are reset as the universe spins toward the future, the higher plane. Therefore, all the material that makes up and surrounds a proton or fourth-dimensional sphere—all its information—is reset instant to instant to the very edges of the world, to that indescribable border of emptiness.

The inertia of shape, the counterforce of gravity with tangible electron clouds and other expanding concentric spheres, is restructured every instant. The information for this inertial rearrangement, this gravitational counterforce, is Einstein's spooky action at a distance. It is what makes up our three-dimensional world.

We sense gravity, and, as Newton stated, it is an instantaneous force caused by every particle attracting every other particle. Inertia, gravity's counterforce, is an immediate resetting of the universe, moment to moment. Gravity and inertia—action and reaction—are the essence of our three dimensions.

But we also see effects between universes, between atoms, leading to motion and change, to movement. By definition, movement can only occur over time, over that imaginary fourth-dimensional distance that must be interpreted temporally. Since any movement takes some set interval to unfold, its signal occurs at a measurable velocity; its limit is the speed of light.

Movements are seen as the electromagnetic wave disruptions emanating from each to every universe, from atom to atom. Thus, electromagnetic emanations are what constitute the fourth dimension. The three-dimensional world is static; it is *what is*. The fourth dimension is motion and change, it is *what is does over time*. Let me try to draw this.

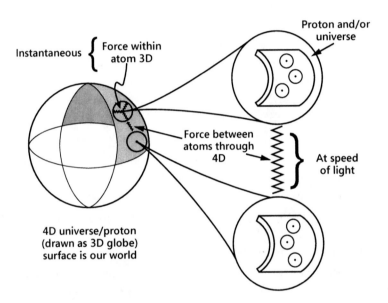

*Question:* Okay, spooky action at a distance, entanglement, is the same as gravity: a force that acts instantaneously to establish our three dimensions. And what we sense as movement is the fourth dimension. But what about elements being visualized as either waves or particles. How do you begin to explain that?

# 21

# WAVES AND PARTICLES

I N THE QUANTUM WORLD, the world of the very small, there are no distinct particles. All entities are at the same time both particulate and wavelike. This
paradox dates back to Newton, in the 1600s, who felt that light consisted of
corpuscles (or particles) because it would not re-establish itself about objects
as would waves.

However, about 1800, Thomas Young performed the two-slit experiment,
which decisively showed that light had to consist of waves (only waves could
generate the interference pattern found). Thus, the wavelike nature of light became accepted theory until Einstein, in the early 1900s, definitively proved
that light also consisted of particles (the photoelectric effect's dislodging of
electrons conclusively revealed that it was composed of discrete objects). Thus,
historically, light has been shown to be, at different times, both particulate
and wavelike.

When the double-slit experiment is performed with electrons, one initially
gets what appear to be particulate changes that, over time, form an interference
pattern; therefore, electrons, or all quanta, are thought to consist, at the same
time, of both waves and particles. What is even harder to understand is that,
if an electron or any entity is visualized, or measured, it becomes exclusively
particulate; it no longer forms waves. Thus, knowledge of the object changes
the outcome of the experiment: no knowledge, waves; knowledge, particles.

## Newton's Corpuscular Theory

*Answer:* Let us attempt to make some sense of wave–particle duality. It is really a problem that dates back to Newton's original concept of light as particulate (corpuscular). He felt that, if light consisted of waves, it would be visible about objects, just as water waves go around obstacles, such as rocks, to reform on the opposite side. Since this did not occur, since light was blocked by opaque barriers, he concluded that light must be composed of particles, not waves.

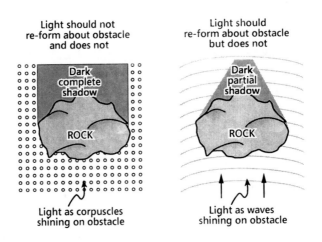

Light should not re-form about obstacle and does not

Light should re-form about obstacle but does not

Dark complete shadow

Dark partial shadow

ROCK

ROCK

Light as corpuscles shining on obstacle

Light as waves shining on obstacle

Therefore, for over 100 years, from Newton's pronouncements until the early 1800s, light was considered to be particulate.

*Question:* But this does not make sense. If the wave length is small and the obstacle large, it is highly unlikely that the wave will reform about that impediment. Take Long Island, and have a wave with a length of about 100 feet (peak to peak) hit it on the south shore. Since Long Island is 120 miles end to end and the wave only 100 feet between crests, it will not be present on the north shore of the island.

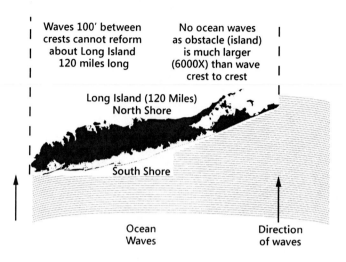

Waves 100' between crests cannot reform about Long Island 120 miles long

No ocean waves as obstacle (island) is much larger (6000X) than wave crest to crest

Long Island (120 Miles) North Shore

South Shore

Ocean Waves

Direction of waves

*Answer:* You are, of course, correct. Newton did not realize that light waves may be very, very small and, therefore, able to only go around and reform about minute obstacles. Thomas Young, however, an English scientist, devised an experiment in the early 1800s that proved light consisted of waves.

## Two-Slit Experiment And Light

In this experiment, Young had a light shine on a barrier with two very narrow slits. He found that the light cast an image of bright and dark lines on a distant wall beyond this barrier—a classic interference pattern. He deduced from this that light had to be a wave. He reasoned that a wave of light struck the barrier, broke up, passed through the slits and reformed as two separate entities, causing this pattern (see illustration on the next page).

*Question:* Well, this is what I already mentioned. If light goes around a small enough barrier (the tiny slits) after the waves break up, they reform on the opposite side, and in this case an interference pattern is visible. It is just like a water wave going around a small rock. It may break up but it will reform on the opposite side.

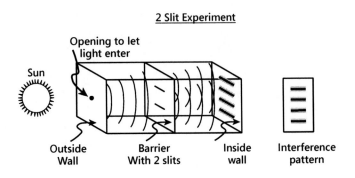

**2 Slit Experiment**

*Answer:* You are absolutely correct. Newton was mistaken; he did not take into account the size factor associated with light. Thus, for the next hundred or so years, light was thought to consist of waves, not particles. Maxwell's theories further cemented this concept, since he proved that all electromagnetic energy consisted of waves moving at the speed of light ($c$).

## Einstein's Photons

However, in the early 1900s Einstein once again upset the applecart. He showed that light, although carried as a wave, also acted as a particle—later named a photon. He explained that, as a particle, light caused electrons to be ejected from material—the photoelectric effect. It was this reasoning that brought Einstein a Noble prize.

From this and other proofs, the current concept of wave–particle duality is derived. Thus, at the present time, it is assumed that photons, electrons, and all other substances exist as both waves and particles in bizarre configurations.

## Two-Slit Experiment And Electrons

Let us return to the two-slit experiment, first done by Thomas Young, using ambient sunlight, and now performed with electrons. When using electrons, the weirdness of wave-particle duality is quite stark.

We will have an electron gun shoot electrons at a barrier with two parallel slits and record the effect on a fluorescent screen. If both slits are open, the fluorescent screen reveals bright and dark lines, fully consistent with Thomas Young's experiment. If we slow the gun so that only a single electron at a time is emitted and, thus, just one spot appears (as that electron hits the screen), after a period an interference pattern is again found. Therefore, even if only one electron passes the two slits and impacts as a particle, it still acts as a wave—it causes an interference pattern.

**Question:** So are you saying that an electron impacts the screen at one spot, and thus is a particle, but over time builds into an interference pattern—therefore presenting as a wave?

**Answer:** Yes, that is what I am trying to show. However, quantum theory is even weirder. If we put a measuring device at each slit and find which one the individual electron passes, the final outcome is not an interference pattern, not a wave, but two clumps, as if the electron remained particulate.

Thus, measuring the electron as it traverses the slits *changes the experiment*. Once measured, the electron persists intact; if unmeasured, it becomes wavelike. No matter how this experiment is performed, and it has been done in countless ways, the same result always occurs.

**Question:** This truly is very weird. You mean that our *measuring* of something—an electron, say—changes the nature of things? Particles become waves or remain as concrete entities dependent on whether we notice them.

**Answer:** Yes, that is what quantum theory has determined. The math behind it is quite difficult, really out of our reach, but the experimental conclusions are always the same. Acknowledgment, measurement, leads to a particulate configuration; lack of the same leads to a wavelike structure.

If a physicist were asked to resolve this paradox, that person would, most likely, declare that the universe is so designed, and that no further explanation is required. The scientist knows the math is correct and the experiments sound, and that the world acts in this mysterious fashion, but likely would not try to explain it any further.

*Question:* If quantum mechanics is so strange as to be essentially unexplainable, then, why bring it up? Why not just accept the weirdness of it and go on as most others do—why even bother?

*Answer:* The reason I bring up the whole concept is that it can be explained and understood from a fourth-dimensional perspective; it is simply in the *three-dimensional* world that visualization is impossible. Let us therefore make the attempt in the following chapter.

# 22

# A 4D PERSPECTIVE

O NE WAY OF VISUALIZING wave–particle duality is to use the analogy of the phases of the Moon. When observing the lunar surface, we see a 2D disc but infer it to be a solid sphere. In a similar fashion, when sensing a quantum entity we appreciate only three of its four dimensions. The constant spinning (in and out of this higher plane) presents "phases" analogous to the Moon's, which are then interpreted as both particulate and wavelike.

In the two-slit experiment, what actually passes from the emitting source (electron gun) to the recording fluorescent screen is the force that is manifested when an electron cloud is initially broken. It passes through any number of slits encountering an ever-present photonic "sea," allowing for waves that, upon impact (at the fluorescent "shore"), register as equivalent to the originally displaced electron. Thus, no particle "flies" through space to contact a screen; it is really a force unleashed by the original electron cloud displacement.

When a device is placed at a slit (to document whether or not an electron has passed), the wave so encountered has been dissociated (from the main force), never to be felt at the fluorescent surface. It is comparable to closing that slit; therefore, when recorded, only a single wave results—there cannot be, nor is there, any interference pattern.

## Wave–Particle Duality

*Answer:* Let us use the phases of the Moon to attempt to understand the wave–particle enigma. We all know that the Moon orbits the Earth—one revolution takes about 29 days (a lunar month). Since the Earth similarly rotates (on its axis), we see the Moon best during the night, when there is no interference from bright sunlight (however, it obviously can also be present during the day). We note different phases dependent upon the extent of the visible illuminated surface (remember, one-half of the Moon must always face and be lit by the Sun). Perhaps a simple drawing would best explain this.

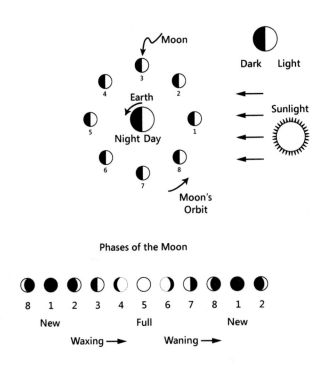

Phases of the Moon

*Question:* So the phases of the Moon are due to its orbiting the Earth approximately once a month; thus, we see slightly different aspects of it each day.

***Answer:*** That is correct. Also we visualize it in flat two dimensions. We see a disc where a sphere exists. Although we know it has three dimensions, we really see only two; we do not see, but only *infer*, depth. Thus, a two-dimensional reflection of three-dimensional reality gives us our well-known phases.

Let us now go to our three-dimensional version of four-dimensional existence. We will find the same *phases* that are present in observing the Moon. We are theorizing that the quantum object is four-dimensional and is constantly spinning in that unknown direction. We can only, however, *understand* it in our mundane lower-dimensional plane (we merely comprehend three of its four dimensions). Therefore, we see a sphere that waxes and wanes, just as the Moon appears to do. Let me draw it.

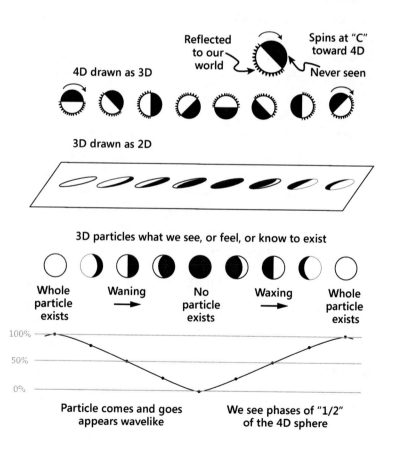

In our drawing the entity continuously changes as it rapidly spins in the higher dimension. We merely see its shadow; thus, we only can glimpse a part, never the entirety. But the variations can be construed as a wave with a frequency based upon its velocity of rotation ($c$—the rate at which the universe moves, the speed of light) and its size (or circumference).

*Question:* Okay, I seem to understand. We, in the lower dimension, can comprehend only that part of an object reflected onto our world. This is exactly analogous to when we view the Moon; we are seeing it in but two dimensions when we know it actually is a solid sphere.

*Answer:* You are correct. To make your comparison clearer, if we were inhabitants of a two-dimensional world, the Moon would be part of our inhabited plane. It would not be distant (in outer space) since there would *be* no such direction (away from our surface). Thus, the Moon actually *would* change shape (two-dimensionally) on a day-to-day basis—all we could comprehend, feel, or interact with would be two-dimensional versions of a three-dimensional realm.

However, we consider our world three-dimensional. We are aware of but three in a universe of four; we cannot understand another direction (it is entirely foreign to our experience). Therefore, objects spinning in the higher plane fully disappear but then miraculously reappear. As we can never see more than one-half (only the outside is visible; the inside is but *potentially* present), we simply glimpse constantly shifting fragments as they grow smaller, disappear, then reappear and enlarge. Thus, to us, they are ever changing (they are both waves and particles).

*Question:* I seem to understand. The wave–particle maze can only be grasped from a loftier level; we are truly lost if we remain on our lesser plane.

## Two-Slit Experiment

*Answer:* You seem to follow my concepts. Now that we have some idea of wave–particle duality, let us try to fathom the subtleties of the double-slit experiment. When done with electrons, if performed one electron at a time, distinct spots at the recording screen are found (particles), which, over time, coalesce into an interference pattern (waves). Thus, the same electrons appear to have a dual reality.

If implemented with recording devices (at the actual slits) seeing which openings the individual electrons traverse, one obtains two distinct clumps; no wave or interference pattern is found. Thus, recording the electrons as they travel makes for particles, and not measuring allows for waves.

What is really occurring when an electron is expelled from some device (gun) and then displayed (on a fluorescent screen) is that an electron cloud (contained within the heated metallic wire of our gun) is disrupted, and an electron is displaced. It is not *shot out* as initially described. The space previously occupied is now open to (unguarded from) the force of attraction of that black hole center, or proton, and this force extends from that gun to our screen.

The force, or pull, as it emanates through the opening of our spinning, disrupted electron cloud, is felt intermittently (each time that dislocated opening passes in front of the proton) through the *barrel* of our gun. This sporadic attraction is registered upon and within a *photonic sea* (situated in the space just adjacent to the fluorescent tube's recording surface) as electromagnetic disturbances or waves. As these photons subsequently impact the screen, an image, equal in magnitude to the original electron's dislodgement, is perceived—in essence, a newly displaced electron.

The screen's recorded image is not caused by a physical electron that has been shot from a gun but by the impact of an attractive force (transmitted via the ever present photonic sea) that disrupts material at the fluorescent surface; and, since the etiology was the initial displacement of an electron (in the gun's heated wire), the result is equivalent (an electron's dislodgement at the screen). Nothing but a force has traveled between gun and target; however, upon impact, waves through our photonic sea (electromagnetic disturbances) result in a spot on the screen where a "supposed" electron has landed.

**Question:** Are you saying that an electron does *not* fly from the gun to the screen? Are you implying that it is solely a *force* of some kind that is established by the *absence* of the electron at the heated wire? Are you then stating that the electromagnetic wave is simply that disturbance through some strange photonic sea that surrounds all atoms?

*Answer:* That is exactly what I am saying. That is why the *phantom* electron can pass through both slits simultaneously. It is not an object moving through space but a force pulling back toward a central source. There never is a particle moving, only a force that interacts with ever-present photons (surrounding all atoms) whose impact allows for a *new* particle's appearance.

Thus, when even a single electron is supposedly sent, it still goes through both slits. This is because the initial weak displacement at the heated wire (gun) disrupts one electron, and the resultant minimal force generated, therefore, merely dislodges the equivalent diminished mass at the screen (only one vague spot). Over time, however, many spots are so recorded and, in due course, describe an interference pattern. (The force, originated by a single electron's displacement, always goes through both slits—any intensity of force will automatically traverse all openings—but its initial strength is so slight as to allow the recording of only a solitary impact.)

## Initial Plunge Into The Photonic Sea

As to the mention of a photonic sea: We have yet to discuss this topic and will in the next few chapters. At present, just visualize a fine particulate sea—a delicate dust—surrounding all material, and when some force (electromagnetic interaction) is registered, disruptions (waves) are transmitted through this substance, which can cause changes when impacting electrons, like a sea wave striking a shore. If of sufficient strength (high enough frequency), these waves can dislodge electrons from the affected material.

*Question:* Okay, I will allow for your cursory description of a photonic sea; however, it needs to be explained more fully in order to really make sense. To get back to what we were discussing, I can now see why a two-slit apparatus should always lead to a wave pattern with interference. But, remember, if one records the slit through which the supposed electron passes, then the final result is not that pattern, it is a particulate clump.

## Recording Devices

*Answer:* You are correct. Recording this passage leads to particulate matter. It is as if, each time an electron is recorded, one-slit, or entrance were closed, hence, merely allowing for a single wave to be registered on the screen.

The only way this could occur is if the detectors at the apertures (during their measurements of our split wave) were to impede what they measure, preventing an impact at the screen. Thus, the act of recording a passage, in essence, closes a slit. It destroys a wave—it washes it up onto the *shore* of the recorder as an electron. That wave cannot reach the screen. It has been intercepted, recorded.

Therefore, the act of recording interferes with the transfer of energy from the gun to the screen. Only a wave that is not measured impacts the screen; the other one is intercepted: It strikes the recorder.

*Question:* Are you saying that, by recording or detecting the electron as it passes the slit, you are dissociating the two waves (one formed at each entrance), forcing one to be intercepted by the recorder while the other strikes the screen?

*Answer:* Yes, that is what I am trying to show. The internal slit recorder is a registering device, just as is the fluorescent screen. Each appears to measure an impact—an electron. However, what they are really evaluating is the force emitted by the displacement of an electron (from the heated wire source) transmitted through space as a wave and finally reassembled at each device as an electron. Both the recorder and the screen are capable of identifying this force, and each establishes it as an electron. But since there are two waves, each recognizes one electron, and, as they are on separate devices, no interference pattern is found.

## Deciphering The Quantum Realm

*Question:* Do you really feel confident of your explanation? It does seem a bit far-fetched.

*Answer:* It may appear strange, but by envisaging the cause of the impact to be only a force, not a supposed particle, a clearer picture emerges. Thus, the answer to the two-slit puzzle is that no particle flies through space, passing two openings simultaneously. Only the force generated by the initial displacement of that particle (the disruption of the electron at the gun) so moves and, as a force, quite easily passes two, or more, slits concurrently.

Therefore, there are at least two reasons for the wave–particle conundrum. They include the waxing and waning of a spinning higher-dimensional object as it presents in a lower realm, and the transfer of a force, not a particle, in the classic two-slit experiment. Each aspect explains a piece of the puzzle: why entities appear, at the same time, both wavelike and particulate.

*Question:* So quantum physics can make sense. Spooky, immediate action is a consequence of gravity's instantaneous velocity, and wave–particle duality is both the visualization of a higher-dimensional object and, in the two-slit experiment, the passage of a force, not a particle.

*Answer:* That, in a nutshell, is what I am trying to show. Quantum concepts can be clarified if viewed from a higher plane. They are utterly confusing when interpreted as occurring solely in three dimensions.

# 23

# THE ILLUSION OF MOTION

A LL MOTION DEMANDS THE 4D, or time, as motion is defined as the change in distance over time. Just as in a movie, each Planck moment is a static frame with minute changes scene to scene. Since we are but aggregates of many, many 3D wave fronts (of 4D entities), as our universe constantly moves forward in time, these "shadows" appear to move; but it is the universe that really moves about them.

Therefore, the world consists of individual frozen instants each lasting one Planck moment—$8 \times 10^{60}$ in total. From a 4D perspective, all these moments exist; there is no time, there is no motion. Thus, from a 4D perspective the future has already happened, it just needs to be "discovered."

## Motion Demands The Fourth Dimension

**Question:** You know, in your previous discussion of wave–particle duality, you spoke of our world as merely the reflection of a higher dimension. But you have also mentioned that our three-dimensional world moves over that great fourth-dimensional sphere at the speed of light, traveling one Planck length each Planck moment. According to you, our universe re-establishes itself each Planck moment, disappears, and reforms the next. This has been going on *forever*, but we can be aware of just less than one quarter of that sphere—13.80 billion years.

However, if our world restructures, each moment, only to disappear and reappear again and again, if we are a great motion picture in three dimensions of fourth-dimensional reality, from where does actual movement arise? In a movie, the individual frames are frozen two-dimensional pictures of a three-dimensional world. Motion, realism, occurs as scenes flit by. Objects appear to move; but movement only happens in the three-dimensional world captured by the two-dimensional images that constantly change.

Where, then, does action or reality come from if we consist only of the static frames, or three-dimensional reflections, of a higher level?

## We Do Not Move, The Universe Does

*Answer:* To answer your question, one must delve a little further into that other plane. Whenever we speak of motion we are discussing the fourth dimension—time—for all movement is but a change of distance over time. Although we think an object moves within the three-dimensional universe, it really is the universe itself that travels into the higher dimension. As you already noted, each third-dimensional moment is a frozen montage, and things are stationary within those moments. The universe journeys through time, and our perspective continually changes instant to instant.

We, as three-dimensional entities, are composed of multiple wave fronts, disturbances caused as the reflections of higher-dimensional objects break and reform. Our very existence, our solidity, is based on these shadows, or projections from another realm. In the real or tangible world, we sense motion and think that three-dimensional items move. However, all are fixed; the universe is a continually restructured monolith that flickers by at light speed on a fourth-dimensional sphere. Each Planck moment, as an old universe disappears, a new one comes into existence to last only that brief instant.

The universe passes through time (into the fourth dimension), and the changing perspective of higher-dimensional entities (the interconnected reflections) are of what we are composed; they contribute to our solidity, to reality. Of course, since we are a part of it, we have great difficulty in full comprehension. Only when we delve into the higher dimension, into the very small (via quantum theory) or very large (by way of astronomical redshifts), do we begin to see all these weird aspects.

**Question:** So motion is an illusion caused by fourth-dimensional objects visualized in three dimensions. There is no motion in our, the lower dimension. The higher dimension is all that exists and we are but a shadow of that realm, a mass of interconnected reflections, yet to us this is everything.

**Answer:** I think that you understand as well as I do. Quantum theory with its weirdness is just an attempt to describe in everyday terms what is really other-worldly. To understand it, one must view things from another perspective.

## The World As Individual Stationary Moments

Let me show you a set of drawings to help visualize the world as I see it. (See next page.) Remember, the universe is really innumerable ($8 \times 10^{60}$) three-dimensional worlds stacked one after another on a great higher-dimensional sphere. It is constantly moving through that dimension, through time, at the speed of light. If we were to take just several of these many, many worlds (in the example, past worlds of 10 billion, 1 million, and 1000 years, then 8 to 9 minutes, and finally 1 to 2 seconds) and display them in very simplified form, we can see how the universe evolves.

Each frame becomes a three-dimensional moment (a Planck interval), and electromagnetic waves travel from one to the next. Because of gravity, they immediately engulf each interval but move, frame-to-frame, at a measurable velocity—the speed of light. Thus, what one *sees,* at any instant, is light, as waves, that have traveled through the fourth dimension (through time). One senses light as moving at a recordable, yet unreachable, pace (the speed of light), but really, within any interval, due to gravity, it is instantaneous.

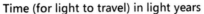

## MACRO 3D MOMENTS

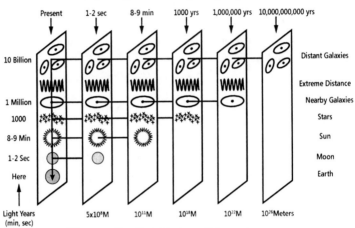

**Question:** I seem to understand. Each Planck interval defines our entire three-dimensional universe. It is immediately reset, moment to moment; however, to *see* it, we must look back through time. Thus, light incorporates each moment instantaneously but travels to the next over time.

## The Future And Predetermination

**Answer:** Yes, remember, from a fourth-dimensional perspective there really is no time or motion; all is set. If all our three-dimensional worlds were placed in front of a higher-dimensional *being,* that entity would perceive the past, present, and future as one continuum. Thus, from a fourth-dimensional perspective, the future has already happened; it is just a sequence of frozen three-dimensional scenes.

The whole concept of free will and predetermination is tied up in the way the universe really unfolds; but allow me to examine that later. Let us get out of the quantum world and discuss some more mundane things—electricity, magnetism, and light. Then we will try to put it all together and show how it affects us day to day.

## SECTION VI

# *Electromagnetism*

HE FOLLOWING CHAPTERS DISCUSS HOW the 4D is appreciated in our 3D world. Electricity is described as the result of a central pulling force and a subsequent counterforce; there is no need for positive or negative nomenclature.

Magnetism is considered to be a rotating vortex formed about a central attracting core. North (N) repelling N and attracting south (S) is due to how these vortices interact. A magnetic monopole, a simple N or S pole entity, cannot exist, as it would require the impossible act of cutting a 4D sphere in half.

Light and other electromagnetic waves are due to the energy felt when a proton's surrounding concentric spheres are disturbed. The actual waves are generated as these instabilities travel through the "sea" of photons distantly orbiting each black hole center. These photons are the fragments of material ejected from the swirling maelstroms of energy surrounding each proton; they are real and have palpable mass.

Although light waves travel at c, photons do not, just as individual water molecules with traversing waves do not. Dark matter, that mysterious material allowing for the rapid rotation of galaxies, is due to these seas.

# 24

# ELECTRICITY

THERE ARE NO POSITIVES, and there are no negatives. These terms are simply semantic aberrations dating back to Benjamin Franklin's initial description of
electricity. All that exists is a proton (the representation of a black hole 4D sphere)
that, in our 3D universe, takes the form of all such 4D entities (spheres within
spheres to the edges of our world).

The electron cloud is simply the initial tangible orb of substance; its existence
is due to the counterforce of inertia. The proton is not "positive" but is surrounded
by a maelstrom of swirling energy that repels other like entities—"positives" repel
"positives." The electron cloud is considered "negative" since it is attracted to this
black hole center; but the attraction is simply due to the centripetal pull emanating
from this core. Protons can be combined (as in the cauldron of the stars), as can
electrons, if their spins are opposed.

The neutron, a supposedly "neutral" object when deep within a nucleus,
soon becomes a "positively" charged proton upon emission; an electron cloud
quickly forms about this new entity. All that has occurred is that the representative of a 4D entity has traveled from the sanctity of a nuclear bed into our
world and now is seen as are all other such bodies—a central pulling sphere
(nucleus) with surrounding, ever-expanding concentric shells to the limits of
infinity.

## Electricity—No Positives, No Negatives

*Question:* You were discussing electromagnetic force as the same as the universal force, or gravity; however, if it is just gravitational force in the fourth dimension, why do we, when speaking of electricity, refer to positives and negatives? Why are protons positive but electrons negative? Why do opposite particles attract but like ones repel?

*Answer:* This is a semantics problem that dates back to Benjamin Franklin, who first established positive and negative nomenclature. It has stuck and muddied the waters ever since.

In reality, the force of gravity caused by fourth-dimensional spheres is all that exists. The cloud of electrons found is the stable counterforce (how such a sphere presents in three dimensions). The force pulling toward the higher dimension (black hole) causes a swirling vortex whose edge repels other like vortices. We call this maelstrom of energy "positive"; thus, positives repel positives.

Obviously, repulsion can be overcome, as all atomic nuclei, other than hydrogen atoms, consist of multiple protons and neutrons. They can be brought together under the immense pressure and heat of fusion in the interior of stars and during supernova explosions.

The vortex caused by the pull toward the fourth-dimensional black hole, represented by our proton, is a vast, roiling whirlpool of energy. Material caught in this quagmire is pulled into the fourth dimension. Some, however, barely missing is spun around and then spewed out into the far reaches of three-dimensional space.

If flung out in a horizontal direction, the material takes on a spiral configuration, similar to our and other galaxies. If expelled in a more vertical direction it stabilizes into a globular shape, again similar to many galaxies. If ejected in a north/south direction, it appears as long filaments emanating from the poles, a configuration also found among galaxies. Let me draw this.

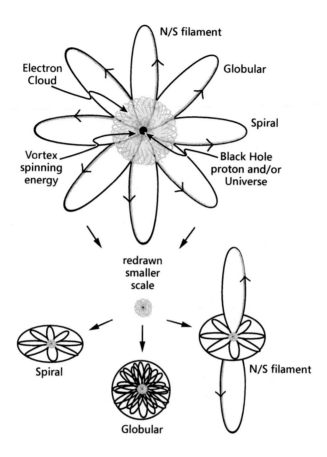

Now, the electron cloud that forms about this gyrating whirlpool of energy is the tangible matter that counters the inward pull. Its rotational speed and distance from the attractive center allow for a centrifugal push equal and opposite to the centripetal pull.

*Question:* But you were saying that there is no meaning to the terms "positive" and "negative," that they are mere verbiage. Why then does an electron have a supposedly negative charge that repels other electrons?

*Answer:* I think that this has to do with the electron's motions—its rotation and intrinsic spin. Although the electron cloud, in order to exist (three-dimensionally), presents as rotating about a higher-dimensional centripetal core, the actual electron itself, whatever it really consists of, has an inherent spin. This spin, however, is not like that of the Earth on its axis but is strictly a hypothetical (or fourth-dimensional) concept.

In quantum mechanics the spin is represented mathematically as plus or minus $1/2$, and two electrons can occupy the same orbit if their signs are opposite; thus, electrons can coexist. Repulsion (electron against electron), therefore, is not secondary to an inherent negative charge but appears to be due to this theoretical dynamic. We will try to come to a more sensible understanding of this in a later chapter; but the concept of "positive" and "negative" is strictly semantics. There are no positives, there are no negatives; there is just the universal pull toward the fourth direction, which we sense as a swirling vortex of energy, a proton, and the centrifugal counterforce, an electron cloud.

## Neutrons—Neutral Charge

Let me delve a little further to help explain this idea. Nuclei (other than those found in simple hydrogen) are composed of both protons and neutrons. Protons are considered positive and are surrounded by equally charged negative electrons. Neutrons, however, are electrically neutral. They are neither positive nor negative. They are not encased by any electrons. However, when a neutron is dislodged from the nucleus, strange things happen.

Soon (within 15 minutes) after the neutron is freed, it changes into a proton surrounded by an electron cloud. This occurs with the release of a neutrino, a weird particle of minimal mass. Why neutrons are usually stable within nuclei but always unstable without is not really explained in physics; the phenomenon just happens.

I am suggesting that, within the nucleus (the portal to the higher dimension), the neutron is so deeply embedded that, in fact, it is attached to the *outside* (fourth-dimensionally speaking) of the facade we call our universe. Our three-dimensional world is established simply by the number and volume of *nearby* protons (objects that make up the *inside*, again in a fourth-dimensional sense, of our surface world). The electron clouds (the three-dimensional manifestations of counterforce), therefore, could only be associated with these particles (protons).

The neutron, while on the outside of our fourth-dimensional sphere, is but a potential body; it would not be interacting with the inner real surface. However, when expelled from the nucleus, when on its own, it would quickly take on all the characteristics of any other higher-dimensional entity presenting in a lower-dimensional realm (a proton with an orbiting electron cloud—our hydrogen atom).

**Question:** Aren't neutrons occasionally unstable, even within nuclei? Isn't there some kind of radiation they can emit when part of a nucleus?

**Answer:** You are much better informed than I thought you were. Yes, there is what is conventionally called *beta decay*. In these cases a neutron spontaneously changes into a proton while still within a nucleus, and an electron cloud is emitted. What is happening in this case would be a shift from the outside to the inside of our three-dimensional surface world, but without leaving the sanctity of the nucleus. Why this takes place is simply unknown. Anyway, in these cases of spontaneous change, the same thing occurs as if the neutron were freely ejected.

Thus, when the neutron is dislodged, it enters our realm representing its own black hole replete with universal attractive force and a vortex of surrounding energy—a positive charge. This vortex repels other like vortices—positive repels positive. At the same time, since it embodies a fourth-dimensional entity, third-dimensional concentric orbs are found about it, out to infinity. The most tangible one is the electron cloud, what we call a negative charge—positive attracts negative.

Therefore, soon after a neutral fourth-dimensional particle enters our world, it becomes a positively charged object with a negatively charged surrounding cloud. In reality, it just represents what any higher-dimensional entity would; we just choose to name things "positive" or "negative" when we discuss electricity.

# 25

# MAGNETISM

MAGNETISM IS DUE TO THE 3D rotation needed to counter the pull toward the black hole center (nucleus). Because of this motion, opposite poles attract and like poles repel.

The rotation in the horizontal axis stabilizes about the equator ($+/- 20^0$ to $30^0$). However, as one travels in the direction of the vertical axis, "substance" attracted toward the center and, just missing, is spewed upward; if stabilized, it orbits in a north/south (N/S) orientation with an east/west twist due to the overall motion. If the rotations are in sync —N/S or S/N—there is an additive effect: attraction. If out of sync—N/N or S/S—there is repulsion.

A magnetic monopole, although theoretically thought to exist, has never been found. This is because all 4D black hole centers, in order to exist in our 3D world, attract and form rotating spheres about themselves. In the vertical axis, a N/S orientation must always occur. If only N or S were to exist, the black hole center would continue to attract all material about it, finally engulfing the entire 3D universe. Thus, a magnetic monopole would lead to the annihilation of our world.

## Magnetism And Spin

Question: Since we were just discussing electricity, why not explore magnetism? Don't electromagnetic waves consist of both electric and magnetic fields?

*Answer:* Yes, all such modalities contain electric and magnetic components, at $90^0$ to each other, moving through space at the speed of light. The magnetic force extends vertically and the electric force horizontally; this is inherent to the rotation seen in three-dimensional entities as they form and stabilize about all higher-dimensional objects.

Let me explain further. Rotational velocity allows for permanence when the centrifugal push counters the centripetal pull. However, since the entire sphere (or electron cloud) rotates essentially as a unit, the velocity as one approaches the north or south pole is much less than at the equator.

Let us use the Earth as our example. Since it rotates once a day, and since its circumference at the equator is about 25,000 miles, it moves 25,000 miles every 24 hours (or at approximately 1,000 miles/hour). However, at the poles its speed is a great deal slower, as its circumference there is markedly reduced (less than one mile verses 25,000 miles). Thus, the velocity of rotation at the equator is around 1,000 miles/hour, but at the poles it is less than 1 mile every 24 hours, a difference of almost 25,000 times. Therefore, as one travels from the equator to the poles this rate keeps decreasing, and with it intrinsic stability.

Now, on the solid surface of the Earth this difference does not affect much except for a slight change in shape at the equator verses the poles. However, on the three-dimensional shell of the electron cloud there is a marked expansion in the middle and a flattening at the top and bottom. The material at the mid-latitudes becomes more stable (solidified) than that at the poles. Substance not rotating at appropriate velocity falls toward the core (just as does the rapidly flowing vortex all about our proton). Most of this is captured by the immense tugging force, but some just misses and is spewed out into three-dimensional space. That motion, in a north/south orientation, is our magnetic field.

*Question:* Since material is constantly orbiting, must there be a direction to this flow? Does it all have to move in but one way?

*Answer:* Yes, all substance must circle about the central pull on the same path; if it did not, there would be collisions and rotation would cease. Using a right-hand rule, the fingers of that hand could be the route of this circular motion and our thumb could then point north. If we were to twist our left hand $180^0$ at the wrist, and position it under our right hand, the southern alignment would now be found. Let me attempt to draw what I have been discussing.

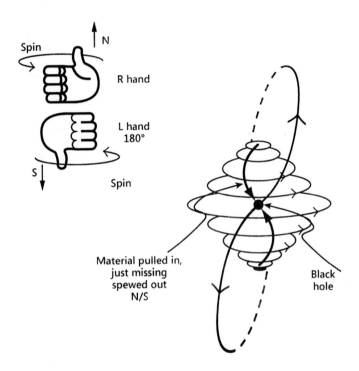

## Magnets—Why Opposites Attract

*Question:* When we talked about electricity, you noted that the terms "positive" and "negative" were just semantic aberrations, that there is really only one force—the force of attraction. But in magnets north attracts south, and south attracts north, and both repel each other. If there is but one force, why are there separate north and south directions?

*Answer:* Again it has to do with orbital circular motion. There are both vertical and horizontal aspects to this motion, best seen as one approaches the poles. The central attraction causes polar material to fall toward it; however, some just misses and is then flung up and out. Thus, we get both a down/up and a twisting circulation (a combination of a north/south and east/west rotation) toward the poles, but a more stable circular rotation toward the equator.

If all our atoms rotate in the same direction, let us say toward the fingers of the right hand, we see opposing motion (out of sync) when like poles meet, but comparable motion (in sync) when opposites join. Thus, opposite attracts whereas like repels. Therefore, magnetism (attraction or repulsion) is due to the rotating shell (our so-called electron cloud) understood in a north/south direction. Let me attempt to draw this.

Magnets

## Magnetic Monopoles

*Question:* Do you recall before, when we discussed the concept of very early cosmic inflation, we spoke of relic particles—magnetic monopoles? Could they possibly exist?

*Answer:* Magnetic monopoles have been theorized to exist yet never found. Whenever one cuts a magnet in half, one always gets both poles. To get just a single pole, one would have to cut the actual fourth-dimensional black hole in half, an impossibility. If it were to be accomplished it would immediately reform. If it did not, this fourth direction would subsume the entire three-dimensional world (pull it all in) and our universe would disappear. Let me draw this concept.

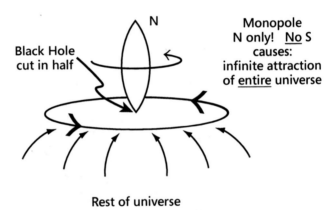

Rest of universe

Thus, a magnet always has a north and a south, and a magnetic monopole cannot exist.

# 26

# LIGHT

P *HOTONS HAVE MASS; THEY ARE fragments formed as material is ejected from the vortices of energy that surround black holes. They are flung outward and orbit in the distant space that surrounds each 4D entity, proton or galaxy.*

Light (electromagnetic) waves are how force is perceived through these distant seas. The force, as it travels through innumerable static 3D universes, is really the same as gravity; however, due to its journey through the 4D (through time), it has a measurable, but unobtainable velocity—c. It leads to disturbances in these photonic seas: light waves.

Although these waves travel at c, the material through which they move does not, just as a water molecule does not when impacted by a wave. The initiating causes of these energy waves are breaks in the concentric spheres of material surrounding protons. As these shells continuously spin, fissures closer to the center (circling at a faster rate) more "frequently" allow uncovering of the attractive core, and subsequent penetrance of greater force. Thus, the higher the "frequency," the greater is its strength.

Dark matter, that mysterious material believed to cause the rapid rotation of galaxies, is due to these photonic seas. Their bulk allows for this extra mass, thought to exist but never found. Finally, electricity (electronic flow) is similar to these waves, only the material itself is different (electrons vs. photons).

## Revisiting The Photonic Sea

*Question:* When discussing the two-slit experiment, you described a so-called "photonic sea." Is it a consequence of the swirling whirlpools of energy, the vortices that exist about each black hole center (or proton) formed by the attraction of material to those cores? You noted that most gets caught, but some, fragments just missing, are flung out to orbit in deep space. Is it this substance that coalesces into your photonic sea?

*Answer:* Yes, that is what I am alluding to. A small portion of these energetic shells escapes and is expelled outward. Its essence reforms (from energy to particulate matter), but, like all solidified energy, it must have palpable mass; it becomes our photons. Photons are not what most believe; they are not particles without substance that move (as *rays* of light) at $c$. They are remnants (of material that had surrounded protons) spewed out to great distances through which energy is now transferred, atom to atom.

All black holes, large and small, are essentially the same; thus, visible galaxies surrounding immense black holes are also merely the ejecta from these swirling vortices. Mixed within and surrounding these galactic structures, there exists a finer material—our photons. This delicate dust has mass and, therefore, cannot move at $c$ (the speed of light). Only the force between nuclei moves at that rate, as $c$ is traveling through fourth-dimensional space. Thus, electromagnetic waves are not constituted of massless particles, but, in essence, are the forces moving through very fine particles of minimal bulk.

*Question:* Are you saying that there are unknown photonic seas extending out from all protons to great distances and that light waves or electromagnetic disturbances of any kind are only waves upon these waters?

## Electromagnetic Waves

*Answer:* That is what I am trying to show. Material expelled congeals and orbits way beyond the electron cloud. Some may be thick and tangible (galactic in nature); however, most is of very fine particulate quality, only thinly dispersed, and becomes our photons. When energy from one distant center is perceived, it is as a wave through this component.

That is why light, or any electromagnetic wave, is not discernible in true outer space. If there are no particles to displace, no wave is found. Once, however, that force encounters particulate substance, waves materialize.

The greater the force encountered, the higher the frequency (or the shorter the wavelength). Thus, the larger the force, the more particles move per second. These vibrations, if of high enough frequency, can cause electrons to be displaced—the "photoelectric effect"; explaining this earned Einstein the Noble Prize.

But these particles are not moving at $c$, it is the energy wave itself traveling at this rate which speeds through them. It is similar to a wave in any material. Take the surface of a pond. The individual water molecules simply bob up and down as the force of the wave moves laterally across the surface. Just place a leaf on the pond and watch its motion. Therefore, when a photon impacts an electron, if that photon were to move in a rapid but perpendicular fashion, it could dislodge that electron (lead to a flow of electricity).

**Question:** Let me see if I understand you. When the force causing light hurtles through empty space, it travels at $c$. When it encounters substance (the photonic sea surrounding all atoms), it causes a wave to form through that sea, and the number of waves per second (the frequency) is dependent on the force or pull back toward what originally caused that light to shine.

*Answer:* I think you comprehend what I am trying to show. Light rays, or electromagnetic waves of any kind, do not propagate through space as ever-expanding wave fronts of massless particles. They move, as gravity does, as instantaneous attractions back toward a fourth-dimensional entity.

However, since the fourth dimension is conceived of as time, these instantaneous forces travel through time, through innumerable momentary and static three-dimensional universes, and their velocity is recordable (slowed to $c$). Upon encountering the extended photonic cloud or sea that surrounds each atom, a wave is formed. This wave radiates forward; we consider it light.

Its velocity is a constant, the velocity of our universe, $c$. It is composed of photons, as this is what is encountered. But the photons move wavelike (up and down in place); they do not move with the light wave, just as water molecules do not move with a water wave.

The force of the wave is directly proportional to the initiating cause, an opening to some fourth-dimensional center. The frequency of this wave becomes its energy, the force that impacts our electrons; the more waves per second, the greater the force.

*Question:* So have we always been wrong about electromagnetic waves? Are they really a pull back toward a source of energy, not a push forward of massless photons?

## Breakage

*Answer:* Yes, electromagnetic waves are secondary to breakages in the surrounding regions of distant protons. The force encountered is proportional to the distance from the center—the closer to the core, the greater the attraction.

The pulling force of every proton universe is, however, dissipated by the material in which it is encased. If this material, for any reason, is rent asunder, the ensuing force felt becomes that much greater. The frequency of the wave formed would be secondary to the rapidity of motion of the intact remnant that encircles our proton, the closer to the center the faster it spins and the more *frequently* an opening or fissure is presented. Since the breach is the main source of attraction, the closer to the nucleus, the higher the frequency and the greater the force.

However, the electromagnetic wave itself is a consequence of a pull back toward a fourth-dimensional black hole center (proton). The force of the wave is felt as the photons strike some electron shore. Therefore, although the light wave consists of photons, and these photons evoke changes upon impact, the light wave is merely the transfer of energy from one center to another—the pull felt by one electron cloud when a distant one has been disturbed. Light appears as a push outward but is the consequence of a pull inward.

## Seeing The Light

*Question:* So if that is true, if all protons have distant surrounding photonic seas, why then do we not see light traveling through these seas? Why do we only see light when it directly affects an object? Why can we not see a wave of light traveling through space from one object to another?

*Answer:* What you see is the result of the impact of the photonic wave upon some electron cloud. You do not see the wave as it moves through the photons. Remember, it is always traveling at $c$; it is never stationary. Therefore, for you to see it occurring, light from it would have to travel to your eye. But light is only found upon abrupt contact; it does not exist as one understands it in empty space. There *is* no light just because photons are disturbed in a wavelike fashion. There is only light when that wave crashes to shore, when it collides with an electron, when it hits your eye.

Thus, light is the motion of a wave of energy through material that has been expelled from the vortex surrounding a proton. All protons are encircled by these whirlpools of energy (all protons repel all other protons).

## Whirlpools Of Energy

*Question:* Where does the energy come from to form these whirlpools?

*Answer:* Remember, the proton represents a fourth-dimensional object; thus, in our world it is seen with surrounding spheres to the edges of our universe. These shells are composed of energy, and the force toward the center continually pulls on them and breaks them apart. The electron cloud is the most stable one; its constancy allows for tangible matter. But the photonic sea is another solidified form of energy, only each photon is much smaller, much less massive than an electron, and is flung a great deal farther out to survive as a distant sea.

If one were to allow our solar system to represent an atom, the very center of the Sun would be the black hole characterized by a proton, the Sun's remaining internal core the swirling whirlpool of energy, its surface the electron cloud, the planets the nearby ejecta, and the encircling distant comets (Oort cloud) the photonic sea.

## Dark Matter

Of course, for this comparison to be correct, all stars would need black hole centers—a very debatable concept. However, a less controversial comparison would be of an atom and a galaxy; in that case the galactic black hole center with its surrounding luminous mass would be the proton (and its maelstrom of energy) contained within an electron cloud. The ejecta are the stars flung from this center. Finally the photonic sea is that very fine particulate material that congeals in the vastness of this galaxy and beyond. It is what dark matter is comprised of and gives extra mass to the galaxy, allowing for the rapidity of its spin.

This helps to explain the *cuspy halo* problem—why theoretically dark matter should be at the center of a galaxy (cusp) but appears to be at its periphery (halo). The reason is that it is composed of the finest ejecta flung the farthest out.

**Question:** So you are equating these photonic seas with the riddle of dark matter, and once you give these photons mass then the puzzling velocity seen in galaxies can be understood.

**Answer:** Yes, dark matter exists. It is simply the photonic seas consisting of very fine particulate mass and in great enough quantity to impart the speed to a galaxy's rotation. If the theoretical value for dark matter is correct, the photonic seas contain five times the mass present in what we consider the visible universe of stars.

## Universal Pull And Its Consequences

*Question:* Let me try to summarize your thoughts. The black hole establishes a centripetal force as its three-dimensional surface rotates at *c* (toward time). This force is felt on the shells of energy all fourth-dimensional entities create in our world. These continuously formed shells swirl about their black hole centers, constantly drawn to them, but some of their energy barely misses being engulfed and is expelled outward. In a galaxy, these ejecta would present as an encircling, rotating crown of stars, in our solar system as the Earth and its sister planets.

However, a great deal is comprised of very fine particulate matter, our photons, and swirls about as a great photonic sea. Energy emanating from other universes causes waves through this substance, our waves of light. This sea, as it has so much material yet is invisible, is what dark matter consists of, and allows for the rapid rotation of galaxies.

*Answer:* You do seem to understand. Light, or electromagnetic waves, are secondary to a pull inward. This is our universal force; it is toward the higher dimension.

*Question:* Well, if that is true for light, what about electricity or the flow of actual electrons? Are electric currents similar to electromagnetic waves?

*Answer:* Yes, the only real difference is that electricity, a flow through electrons, is usually seen in tangible material, not in the thinly dispersed photonic sea. It is due to the same force and, therefore, moves in a like fashion; however, just as in the photonic sea, the individual electrons (constituting the electronic sea) act as previously described water molecules and merely move in place (up and down forming waves); they do not travel with the flow.

Thus, all the manifestations of the universal force—electricity and electromagnetic waves—are the same; they are merely a consequence of this fourth-dimensional pull.

# 27

# MASS

OST 3D ENTITIES ARE SHADOWS *of 4D reality. Thus, almost all particulate substance exhibits waves (wave–particle duality, the way we visualize a higher dimension in our 3D world). Electromagnetic or light waves, however, are a different manner of wave. They are the energy in our universe transmitted from one central core to another through photonic or electronic "seas." They are not due to the geometry of 4D objects.*

*Photons are the barely surviving fragments of vortices (which have been consumed by central core nuclei) and are found orbiting in clouds far from these cores. They are unlike other particles; as being solely remnants of 3D matter (vortices), they themselves are not true reflections of 4D entities. Thus, they do not partake in the wave–particle enigma.*

*Finally, although mass (congealed energy) cannot travel (in 3D space) at "c," it can so move as part of the entire universe, continuously spinning toward that higher plane (toward time). Mass is but the depiction of a higher essence; it is the shadow or reflection of 4D existence.*

## Return To Waves And Particles

**Question:** Before, when discussing the weirdness of wave-particle duality, you said that it was due to how we visualize spin through a higher plane. All fourth-dimensional objects exhibit this enigmatic duality, as all move at $c$ in that unknowable direction. Thus, do all entities present with this unusual characteristic?

*Answer:* Let me try to explain my concept. Most matter is but a representation of a higher-dimensional reality; therefore, when visualized in our world, protons, electrons, just about all other submicroscopic objects, have to present as particulate and wavelike at the same time. Thus, they must, when observed, reveal a continuously changing wave pattern. But these waves are not similar to light waves; they are not the energy transferred from one central core to another. They are simply waves formed by the geometry of how we comprehend a fourth-dimensional object.

## Photons—Remnant Particles

*Question:* What about photons? You describe them as the delicate dust surrounding all black holes (protons and galactic cores) formed by the remnants of energy shells that graze but just miss these centers of attraction. Therefore, they are not truly fourth-dimensional objects; they are but the remains of the third-dimensional shadow world (orbs that formed in our universe to represent the higher sphere in our lower dimension). Thus, if they are merely what is left (when these orbs are pulled into the all engulfing black hole), they are not truly fourth-dimensional and should not take part in the wave–particle paradox.

*Answer:* You bring up a very interesting point. The photon is a remnant particle; it is not a shadow of a higher form but a segment of that shadow. Thus, it is not truly a representative of a fourth-dimensional sphere and, therefore, should not be simultaneously wavelike and particulate. It is a special type of object, only a small portion of a third-dimensional entity and, hence, wholly of that plane. It would, therefore, not be part of the wave–particle enigma observed in other material.

*Question:* So the photon, besides being the missing mass of dark matter, is composed of an unusual kind of material. It is a singular case; it is entirely three-dimensional.

*Answer:* Yes, that is what I am trying to say. The photon, unlike other particles (or fermions—electrons, for example), is simply a third-dimensional construct and, as we will see later, due to this it does not have a distinct opposite or antiparticle, and one full spin, or rotation, is comleted in $360^0$.

*Question:* You always state that our world moves at $c$ through the fourth dimension. But objects that exist must also have mass, and anything with mass cannot reach the velocity of light. Don't you see this as a contradiction?

## Mass Is 3D; Awareness Is 4D

*Answer:* Let me try to explain. Once something has mass, it is a shadow (or a shadow's remnant) of a fourth-dimensional construct; it can no longer move through what we consider the three physical dimensions of space at $c$. Only if it were massless could it so move. But only waves of energy are massless, and only such waves move at $c$; most particles are fourth-dimensional, all have mass, and therefore travel at a slower pace.

What we consider our three-dimensional universe is really an abstraction; it is how we understand things. We can only *sense* energy waves. We can only *see* light emanating from some central core through photonic seas impacting our eyes. We can only *feel* emanations from these cores vibrating through our photonic seas. All we are aware of is the third-dimensional surface world, the vibrations through these *waters*.

Thus, our awareness is only of three dimensions. We naturally think, therefore, that our world is solely three-dimensional; we see and feel it to be so. But really it is an extension of a higher realm, we are simply not designed to sense that hidden plane. It is only as we have peered outwardly with ever greater telescopes, or inwardly with ever more powerful instruments, that the fourth dimension has become apparent.

We attempt to explain things three-dimensionally. We consider the universe *expanding*, we think of quantum actions as *weird*, but these misconceptions are due to our trying to describe a higher form in our terms. Once we understand that a superior plane is within and about us, a more sensible concept is possible, an explanation that incorporates that all-encompassing dimension.

**Question:** Are you saying that the three-dimensional world is but a subtle reflection, a thing that lasts momentarily each Planck instant, only to reform again and again? Is it but the motion of the waves of energy lapping onto the innumerable electron shores?

**Answer:** That is what I am trying to show. The three-dimensional world is evanescent, it is an abstraction, it is how we sense the fourth dimension. The true reality is the higher plane. We are, however, as are most other things, designed to but take note of, to inhabit, the lower realm.

**Question:** But if our three-dimensional surface world, moving about a great fourth-dimensional sphere at $c$, is only a representation of a higher-dimensional reality, how can it exert force? Remember, you explained gravitational force as centripetal force and showed that it equaled $mc^2$. However, this was because the entire universe with all its mass traveled in a fourth-dimensional direction at $c$; but we know that only energy, not mass, is capable of such movement. How can, then, our third-dimensional skin, our universe, exert its force if it has mass and, therefore, cannot move at this velocity?

## Protons—Shadows Of Other Worlds

**Answer:** Each proton represents an entire universe. We obviously cannot get innumerable, equally massive worlds to move with ours. However, we can get the shadows of these worlds to so move. Thus, we have the appearance of many, many universes traveling toward a fourth dimension as we so journey, and the velocity is that of our universe toward that dimension—the speed of light.

Try not to mistake the movement of our entire world as a movement through the three physical dimensions of space. It is really the motion toward a higher dimension, toward time, and with it the representations, or shadows, of all other entities. The direction of the universe moving at $c$, therefore, is not through what we understand as physical space but through time; thus, mass in our three dimensions can move at $c$ as we all travel toward the fourth, the future.

Remember, all objects are intertwined; all exist within all others. Thus, each universe contains every other universe. All move as one through the fourth dimension, through time. But each is only represented, is only a shadow within the others, and it is that reflection we equate with mass.

**Question:** So mass is merely a depiction of a higher reality. It is how we conceive of a fourth-dimensional entity. It is how we view the proton universes that make up our world. Each is a universe unto itself. Each really has the mass of our entire universe; however, what we sense is but its shadow. Thus, the entity's mass is simply a reflection, just the tiniest fraction, of its real bulk.

Therefore, the force of gravity, the centripetal force of our universe moving about a great fourth-dimensional sphere toward time, the great inwardly directed force that is the basis of our universe, is based on an illusion, a shadow, the remnant of a much greater actuality. It is how we comprehend an unknowable direction, how we are made aware of that dimension in what we conceptualize as a three-dimensional world.

**Answer:** Yes, the only reality is fourth-dimensional. We understand it, however, in three-dimensional terms; we sense the electromagnetic emanations of swirling, broken spheres about distant protons as three-dimensional waves of energy, as light. But this is simply the surface that we can fathom. The true depth of what exists is within the higher plane.

# SECTION VII

# *Force And Counterforce*

The central pulling or centripetal force $(F = mv^2/r)$ causes gravity and also allows for all energy. It is equal to energy in an unknown direction, within all that exists. Thus, all energy must be constant; it must be conserved.

Inertia is explained as the counterforce that allows for 3D substance, the force equal and opposite to gravity. Just like gravity, it too is instantaneously felt throughout our universe.

Einstein's special theory of relativity is discussed as an interpretation of our 3D world in a greater 4D reality. A 2D example is used in which height, an unknowable direction into a 3D "super" universe, is shown to be one Planck length to that higher dimension. Movement through that 2D universe, layer to layer, would constitute its reality.

In our world, changes in length and time are how we perceive a higher reality. All that exists is the 4D; however, our awareness is of its shadow world—of 3D waves.

# 28

# FORCE—ENERGY TOWARD THE CENTER

THE FIRST LAW OF THERMODYNAMICS is conservation of energy. This is based on centripetal force—gravity—as the source of all energy. If:

$$F = mv^2/r,$$
and m is the world's mass,
and v its speed of rotation c,
and r is imaginary (i) or 4D, then,
$$F = mc^2/(i), \text{ or}$$
$$(i)F = mc^2 = E.$$

Thus, all the energy of the universe, $mc^2$, is really just gravitational force directed toward the fourth dimension—(i)F. This is a constant and cannot increase or decrease, it can just change form.

Inertia, seen, for example, as the sudden jolt felt when a subway starts, is the counterforce to the continual pull toward the center—to gravity. It is what we are composed of. It is what makes up our 3D world.

Finally, the basis of electromagnetic force to mankind for millennia was the light and warmth of the Sun. It is only in the last 200 or so years that we have had any inkling as to the real nature of this force.

## Conservation Of Energy

*Question:* You have been discussing quantum theory in fourth-dimensional terms; spooky (instantaneous) action and wave–particle duality, you feel, are both due to this imaginary but real direction. You state that electricity, magnetism, and light all present as reactions, or counterforces to this direction, and dark matter is its fragmentary (photonic) residue. You also note that, whereas electromagnetism has a measurable but unobtainable velocity (light speed), gravity is instantaneous, yet both are the same; one traverses the fourth, the other the third, dimension. Everything you mentioned, so far, involves this higher plane.

What about energy itself? Haven't we always been taught that energy must be conserved, that none can be added or lost, just altered in form? Does this also stem from what you have been discussing?

*Answer:* You are describing the first law of thermodynamics, and it certainly can be explained from the perspective of a higher dimension. Remember, all the energy of the universe is really derived from the centripetal force ($mv^2/r$) of gravity. Let me once again show you this; since,

$$F = mv^2/r, if$$

$m$ is the mass of the universe, and if
$v$ is the speed of its rotation $c$, and if
$r$ is an imaginary distance ($i$), then

$$F = mc^2/(i), or$$
$$(i)F = mc^2 = Energy.$$

Since $(i)F$ is gravity (a pull toward a fourth or inner dimension), it and the energy of the universe ($mc^2$) are the same. Therefore, all the energy of the world is really but the attraction toward a higher or imaginary direction; all the energy we sense is secondary to the pull caused by the constant fourth-dimensional rotation of the black hole surface, our home. Since this motion is fixed, so too is the total energy of our universe. Hence, energy cannot be gained or lost, it can only change form.

*Question:* But when you spoke of the sea of photons (particles that are flung outward to distantly orbit each black hole, each proton), you said that they were simply congealed remnants of energy spheres (pulled toward these centers, narrowly missing and then hurled deep into space). If that is true, as these energy spheres are constantly reformed, are you not, then, creating *new* energy; are you not breaking a universal law?

*Answer:* I do not think so; most, nearly all, of the actual energy sphere is captured by the black hole itself, only a minute amount just misses annihilation (to be spewed outward into far-away orbit). Therefore, when a surrounding shell reforms due to fourth-dimensional geometry, it is simply re-establishing what previously existed. The minimal amount ejected should be balanced by equal substance falling from orbit. For everything created, a similar quantity would be lost. The only force is the universal pull toward the center; however, the three-dimensional world exists as a counter to this force—it is there because of inertia.

## Inertia

Ernst Mach, a theoretical physicist who greatly influenced Einstein, would say that the jerk felt when a subway started was caused by the mass of all the stars in the heavens acting on one's body. What he was alluding to was that each particle in the universe interacted with every other particle. Thus, each atom, each rotating fourth-dimensional sphere with its surrounding three-dimensional tangible and then barely tenuous shells, is constantly re-established with all the other similar fourth-dimensional entities.

Inertia, the sudden jolt felt when motion begins or the continuing in motion our bodies sense, the reason for seat belts when slamming on the brakes, is the way this interconnected force is felt. It is instantaneous and the inverse, the direct opposite, of the universal attraction—gravity. Remember, Newton stated that every action must have an equal and opposite reaction. Thus, inertia is the reaction countering the force exerted by all other distant particles.

Inertia is the counterforce against which gravity establishes the universe. It is the centrifugal force of the Earth moving about the Sun or an electron cloud about a proton. Without it, fourth-dimensional entities would engulf all existent mass. We would not be here. Inertia is what allows for third-dimensional reality.

## Fire And Electricity

*Question:* You always equate all forces; is inertia, then, simply the universal counterforce?

*Answer:* Yes, that is what I am trying to show. Since the electron cloud is the result of inertia (the pull back of any substance against the universal force) as already noted there really are no positives or negatives. Remember, this nomenclature only dates to Benjamin Franklin. He first discovered that lightning and electricity were one and the same. His concept of a *flow* was the initial reason for the terminology, and the names have stuck; however, they have no intrinsic meaning. Electricity and electromagnetism need no pluses, no minuses; they are merely the effects of an overwhelming attraction toward a higher plane.

Humans have only recently, for the last one to two hundred years, been really using electricity. The concept has merely been understood since the writings of Maxwell. However, prior to that, we employed electromagnetic force; we simply called it fire. The chaotic breakage of electron clouds, initiated by sparks flying off a flint, led to combustion, to flames.

Massive conflagrations, huge forest fires caused by lightning, were early humans' most significant encounters with electromagnetic force. Then, humans discovered how to start and control these wild blazes. Only very recently have we been managing this force, allowing it to do mechanical work or to transmit information. What we understood for eons about electromagnetic force was simply that the Sun existed and that fires burned. There was never a need for positives and negatives, and there still is none.

The Sun, with its wonderful radiant light and warmth, giving rise to life, is our most important encounter with electromagnetic force. We have tamed this force somewhat: We use fires to warm ourselves, to propel machines, and to generate electricity. But all electricity really consists of is just controlled, chaotic combustion. The electrons are still broken off by a pull toward fourth-dimensional oblivion just as in prehistoric blazes, but more coherently (more wavelike), and the force freed can be routed through wires or space and employed by many marvelous machines.

We consider ourselves civilized; we have conquered this electromagnetic demon. We use the energy in newer and more important ways; however, it still is part of that immutable flame (the force of gravity, the universal attraction), only now directed for productive use after its skillful creation by our breaching of electron clouds.

# 29

# VELOCITY AND CHANGE

G RAVITY, DISCUSSED AS THE EQUIVALENT *of centripetal force in a 4D direction, is seen in our 3D world as what mass "does." All mass is made up of atoms whose nuclei are black holes pulling toward their centers; thus, the greater the mass, the greater the number of 4D centers and the greater the gravity. This force is weakened by distance, diluted by the area over which it is spread; it is, as Newton so succinctly stated, equal to mass divided by distance squared.*

*Although the velocity of gravity is instantaneous (2.4 x $10^{69}$ m/s) and the speed of light measurable (3 x $10^{8}$ m/s) they are both equal, as gravity is 3D and light is 4D; they simply are separated by that ever-present constant 8 x $10^{60}$. Therefore, distance in 4D is less than in 3D by 8 x $10^{60}$, but time in 4D is greater than in 3D by 8 x $10^{60}$.*

## Gravity—What Mass Does

**Question:** Let us get back to your discussion of universal force, of gravity and electromagnetism. You have noted that the universe does not move smoothly but only appears to. There is a constant stop and reformation every instant; things constantly remake themselves. Gravity, then, is felt throughout our re-established universe moment to moment; therefore, it must be so rapid as to appear instantaneous.

The electromagnetic force, on the other hand, causing waves of light, a separate manifestation of the universal attraction, occurs between distinct tenuous universes—within the fourth dimension. As we cannot visualize this foreign spatial direction, movement through it is sensed as time. Thus, gravity is within our three dimensions and is instantaneous, but electromagnetic force is outside our three dimensions, subsumed by the fourth, and takes time.

*Answer:* You seem to get my ideas. Remember, gravity can be understood as a centripetal pull established by rotation toward a higher dimension (at the speed of light) of all the mass in our universe. All energy can be explained as secondary to this motion. But gravity is also considered, in the real world, to be due to the mass of any given object. It is thought to be what mass does, the force that any given particle exerts on all other particles, and it is felt instantaneously.

Thus, although it is the overall force in our universe, and holds our universe together, it is also sensed and caused by all separate objects in what we consider our three-dimensional world. Each entity pulls every other one toward its core, toward that imaginary fourth direction. Since most things are composed of individual atoms, and since all atoms have central nuclei, the force exerted (by each on all) is the sum of the attractions of individual nuclei out to the edges of our universe.

The greater the number of such nuclei (of such fourth-dimensional black hole spheres), the more pronounced is the inward pulling force; thus, gravity is proportional to mass. Also, as one goes from the center of each nucleus, the attraction weakens dependent on the distance traveled. Therefore, gravity diminishes (the force is spread out) consistent with how far away from that center it is measured.

Since gravity is equally diluted over the entire expanse, in three-dimensional terms its force would be represented by the exterior of a sphere (an entity with all points equidistant), and the decrease in strength would be based on that sphere's surface ($4\pi r^2$)—or proportional to the square of the radius. Therefore, one can say it is diminished in strength by its distance away squared.

This is what Newton succinctly stated. The strength of gravity is based on the mass of an object and reduced by the square of its distance.

Since all nuclei or black hole universes stretch out to the edges of our universe, each encompassing all others, the force from each is encountered no matter how far one travels only in exponentially lesser and lesser amounts. So the force is always there, even if practically immeasurable, and it cannot be blocked by any intervening substance. It encompasses all.

## Unity Of Velocity

*Question:* I seem to understand what you are alluding to with gravity and electromagnetic force as one and the same. However, it still is difficult to understand how they can be similar yet move at such different rates, instantaneous verses the speed of light.

*Answer:* This is the perplexing problem that constantly occurs when matching third- and fourth-dimensional concepts. It always brings up that annoying number, $8 \times 10^{60}$, that ratio that divides the third and the fourth dimensions.

Let me give you another way of looking at this problem. Remember, much earlier we discussed Einstein's approach to the velocity of light phenomenon. He just accepted the fact that the rate was constant no matter at what speed one traveled. Because of that, as one sped up, length diminished and time expanded; of course, this was only noticeable when moving very, very rapidly.

I showed you the math *trick* that allowed one to grasp the alterations in length and time that accompanied the velocity changes. All we did was assume that all velocities with regard to light were the same, as the speed of light was constant no matter how fast one traveled. Then the distortions in distance and time made sense.

Let us use the same approach now to an instantaneous velocity verses the speed of light. We are stating that velocities in the third and fourth dimensions are the same. In the third dimension the velocity allows for immediate action, it is the fastest movement conceivable. Therefore, the entire distance of the universe ($1.3 \times 10^{26}$ m) is traveled in the least time possible, a Planck instant ($5.4 \times 10^{-44}$ s). This velocity is $2.4 \times 10^{69}$ m/s.

Thus, if 3D velocity = 4D velocity, then:

$$2.4 \times 10^{69} \; m/s = 3.0 \times 10^8 \; m/s.$$

Now if we use the factor ($8 \times 10^{60}$) then the math is:

$$2.4 \times 10^{69} \; m/s = 3.0 \times 10^8 \; m/s \times (8 \times 10^{60}).$$

Therefore, meters, or any measure of distance in the third dimension, are $8 \times 10^{60}$ times as great as they are in the fourth. To make them equal, one has to multiply by that number, thus, the fourth dimension is $8 \times 10^{60}$ (eight trillion, trillion, trillion, trillion, trillion) times as small as the third. The same equation can be made equal if we were to do the following:

$$1/(8 \times 10^{60}) \times 2.4 \times 10^{69} \; m/s = 3.0 \times 10^8 \; m/s.$$

Now we multiplied the seconds in the third dimension by $8 \times 10^{60}$ and, hence, seconds, or any measure of time, are much smaller in the third as verses the fourth dimension.

I am showing that, if the velocity of gravity (instantaneous) is the same as the speed of light (constant, measurable, yet unobtainable by any material object), then distortions between the third and fourth dimensions are found in distance and time.

*Question:* I see what you are saying. The speed of light, the limit in our world, is really instantaneous. Thus, it is actually the same as gravity, except one is a measure in a higher, the other in a lower plane.

## Distortions In Length And Time

*Answer:* Right. We are on the surface of a fourth-dimensional rotating sphere moving at the speed of light but toward the fourth dimension or direction. We cannot visualize this direction. All we are aware of is time.

However, when we look far enough out into our universe, when we observe what has transpired in the distant past, we see redshift distortions and can infer a higher dimension. We sense the motion of the universe as the speed of light, but this limiting rate—this constant—is really the same as the instantaneous velocity of gravity.

*Question:* So when things travel faster and faster, even when approaching the speed of light, light always travels at the same speed compared to them. You are saying that this is because it is really a fourth-dimensional version of third-dimensional instantaneous velocity.

*Answer:* Yes, the distortions noted by Einstein, his special theory of relativity, would be secondary to fourth-dimensional objects coming closer to our three-dimensional plane, thus contracting in length, expanding in time, and becoming more massive—revealing their true fourth-dimensionality—the faster they move.

# 30

## EINSTEIN REVISITED

OUR AWARENESS IS 3D, but the universe is 4D. There are 8 x 10⁶⁰ static 3D worlds making up our universe, and each 3D world is but one Planck length in 4D thickness. To help one visualize this, a 2D example can be used.

In such a world there would be no height. However, if that world were to exist in a 3D "super" universe, it would need some actual substance or thickness. The least amount possible would be one Planck length but in a direction inconceivable to its inhabitants, toward an unknown dimension. As in our world, movement, layer to layer, through this dimension would constitute reality.

Einstein's initial, or special, theory of relativity is based on how we interpret our 3D world in such a 4D reality. It essentially shows how a 4D object's shadow changes as it approaches our 3D plane of awareness.

### Einstein's Initial Theory

**Question:** This appears to be an appropriate moment to return to Einstein's initial ideas of length and time distortions based on velocity changes. Do you want to make any additional comments concerning his original theory?

**Answer:** Einstein's findings of length contraction and time dilation as velocity increases are really based on the higher dimension. Due to this real (seen in the redshifts of distant space) but to us seemingly imaginary direction, the faster things move the closer to our plane they appear. Distances contract, finally reaching the ultimate size (one Planck length), but time likewise expands by that same factor ($8 \times 10^{60}$).

We only visualize three dimensions, but motion demands a fourth—to move we must go from place to place and that can only be accomplished over time (our conception of the fourth dimension). The universe is really a series of static three-dimensional sites constantly progressing toward an unknowable or fourth direction. Movement is only in this direction, and since it differs from our perspective by a factor of $8 \times 10^{60}$, size diminishes but time expands, minutely at slow velocities but measurably at rapid ones.

Remember, there are $8 \times 10^{60}$ Planck lengths or moments that make up our universe, and if, each moment, a new universe, in all its glory, is formed, this universe to a fourth-dimensional entity would be but one Planck length in size (when measured in that dimension). Since we can only understand things in three-dimensional terms, what is actually instantaneous (gravity) is sensed as occurring over time (as light) with its velocity diluted by the number of universes it must encounter ($8 \times 10^{60}$). Therefore, Einstein's findings of distortion in space and time are really because light travels at a velocity that appears to be measurable, but is actually instantaneous as it traverses the fourth dimension.

*Question:* It is still hard to comprehend your concept of a full universe being only one Planck length in size. How can one better picture or understand this?

## Peeling An Onion

*Answer:* Probably the easiest way to see this is to re-imagine a two-dimensional world. If our three-dimensional world were really only two-dimensional, then there would be no concept of height. All would be on a surface, such as this page. Another page above or below would be inconceivable.

But no matter how thin this two-dimensional world appears, to the three-dimensional *super* universe it must have some thickness. All things that exist must have substance. The light cast on a wall, a true two-dimensional world, must, to be real, have some texture. Therefore, existence to a higher dimension connotes some thickness in that lower plane. Since this would be the least possible, it would be one Planck length; nothing real can be smaller.

Thus, to the three-dimensional super world, our postulated two-dimensional universe would have real depth, one Planck length. To us, denizens of that flat realm, there would be no such thing; we could not picture it. But to those on the higher plane, our universe would consist of many layers neatly stacked together, $8 \times 10^{60}$ to be exact; it would be somewhat similar to an onion.

*Question:* So if we were only two-dimensional, we could not conceive of height, which would be the smallest size possible, one Planck length. Therefore, are you saying that, since we are really three-dimensional, our distance into the fourth dimension is that same smallest amount possible?

## Traveling Through The Layers

*Answer:* Yes, the distance toward the fourth dimension is the least possible, one Planck length. To us, it is tiny, but from a different dimension it is our entire universe. They are stacked one on top of another like layers of an onion, only we are incapable of seeing that; we cannot sense distance into the higher dimension just as two-dimensional beings cannot comprehend height. But we travel through them, always in an unknowable direction, at a velocity we consider the speed of light. This rate is constant but unreachable (the same as the instantaneous velocity of gravity); however, our journey extends over a much greater distance ($8 \times 10^{60}$ times as far); thus, it is that much slower (it takes $8 \times 10^{60}$ times as long).

Einstein understood this as a change in length and time. What it really means is that an entity as it gains velocity approaches our three-dimensional surface of awareness (we visualize its fourth-dimensional properties more clearly). The closer it gets, the smaller and more massive it would appear. Thus, Einstein's initial theory can best be understood as a fourth-dimensional construct.

## Closing The Circle

*Question:* So you are finally coming full circle. You are explaining Einstein's original or special theory of relativity, with its length and time distortions caused by changes in velocity, as a fourth-dimensional concept. Therefore, when he explained his theory based on the constancy of the speed of light, he was really saying that, since light (or electromagnetic force) is only a manifestation of gravity, and gravity is instantaneous, light is really an instantaneous force but spread over the entire four dimensions of our universe—hence measurable but not obtainable.

*Answer:* Yes, we have come full circle, so to speak, and yes, that is how one can visualize Einstein's theory. The math trick we initially used (all velocities are the same when compared to that of light), to show the distortions in length and time, now is understood as explaining a fourth direction—real (as in outer space) yet imaginary (as in our ability to sense it). It is no trick but the way the real world is shaped verses the way it is perceived, the fourth verses the third dimension.

# 31

# A HIGHER VIEW

T HE UNIVERSE CAN BE SEEN *as a motion picture with static 3D frames contin-uously moving through time. However, we cannot view this movie, since we are a part of it. The basis of all is 4D reality; we are but shadows of this reality, in a sense, figments in the mind of God.*

## The Universe As A Movie

**Question:** Do you recall our discussion of a motion picture—each frame is only a two-dimensional depiction of reality, but continuous frame-to-frame motion creates the appearance of movement?

The continuous re-establishment of three-dimensional space, the constant jittering that occurs as each Planck instant fades into the next, is quite similar. Therefore, the three dimensions of a Planck interval would be analogous to each motion picture scene, and, since there is no movement in any individual one, movement would be due solely to the repetitive change—Planck moment to moment.

**Answer:** I like the analogy; however, the difference is that when we watch a motion picture we are three-dimensional entities watching two-dimensional scenes flick by. In the stop-and-go of the universe, Planck instant to Planck instant, the universe is three-dimensional but we are not *separate* from it, we are a *part* of it. Thus, we cannot view the movie of the universe, we *are* that movie. The three-dimensional instants are merged into each other, and an incessant movement of all things is seen, and it is us.

Motion, therefore, is always from one frame to another, almost $10^{44}$ times a second; but usually one's movement is so slow that no distortion occurs. It has only been in the last fifty, or so, years that we have been able to travel fast enough (rockets, space flight) so that even minute distortions can now be recorded. All motion, all action is, thus, frame to frame; each but the frozen montage of an instant. However, the continual motion of these scenes allows for movement, for the real world.

The fourth dimension gives rise to what is, to all that exists. It is the basis of all force and movement. It is real but at the same time imaginary. All objects reside in four dimensions but are perceived as shadows in three. The faster they move, however, the closer they come to our plane of reality—our comprehensible world.

## The Mind Of God

**Question:** So where does everything come from? Why does anything really exist?

**Answer:** The entirety, the individual electron clouds with their adjacent tenuous shells extending to the very edges of our universe, is how our world shapes itself to counter the pull toward the black hole core. It is the inertia of three dimensions, the counterforce of substance that gives us form.

Our universe is a three-dimensional rotating surface of a fourth-dimensional sphere. Gravity is a consequence of this motion. But for three-dimensional existence to be real, inertia, the reaction to gravity, must also be present. We are merely the product of this inertial counterforce, equal and opposite to the universal pull.

All we consist of is three-dimensional *material* superimposed on a fourth-dimensional *reality*. We sense force and movement (due to the geometry of the fourth dimension); our three-dimensional world (established each Planck instant) is only the counterforce.

We are three-dimensional; we are the inertia to the attraction of the higher plane. All movement can only be toward that direction. All there is, is the fourth dimension—all there is, is God. We are but the inertia of that dimension, figments in the mind of God.

# Part Three

## REAPPRAISAL

THE THIRD PART OF THIS BOOK begins by reviewing all that has been discussed previously. We summarize the concept of the ultimate cause, a higher dimension, seen in all things, and how it affects the physical world. We then examine current beliefs showing how basing all on a fallacy leads to a multitude of mistakes.

The fundamental problem with accepted theory is the assumption of an expanding, hence cooling, universe where particles, essentially, condense out from a primeval cauldron. However, as there is no expansion, no cooling, all of established theory built on this crucial misconception falters. A more logical concept based on a higher dimension is shown to more readily explain our world.

# SECTION VIII

## *Summarize And Criticize*

T HE MOST FUNDAMENTAL FINDING IN CONTEMPORARY ASTRONOMY—an increase of redshift with distance—is reinterpreted as due to a higher-dimensional curve; thus, there is no expanding universe. Given this paradigmatic shift, and assuming a universality of all force, a clearer understanding of electromagnetism and quantum mechanics is obtained. Other concepts are also found to make more sense: baryogenesis, virtual particles, the large-number hypothesis, and length/time distortions.

Current particle theory, the Standard Model, is erroneously based on a primeval explosion—the Big Bang—leading to expansion, cooling, and breaking of symmetry. Although the underlying math may be correct, the initial premise is not, and the conclusions, therefore, are illusory.

# 32

# REVIEW

THERE IS AN UNDERLYING FORCE *that constitutes the universe. People of faith consider it God; those of science, gravity. This force leads to existence; it is sensed by humans as a constant need of accumulation and acquisition, a hunger for more. Societies become ever more efficient under its aegis. Humans over-accumulate, leading to vascular changes and "disease." Free time, a consequence of this efficiency, leads to diversions, the most significant of which is religion. Religion (or spirituality), as a contemplation of this infinite essence, often satisfies the needs brought on by civilization—it becomes a stabilizing force.*

*Einstein searched for this unifying concept; his theory of relativity revealed alterations in space and time that are best understood as due to a higher dimension. Hubble saw these distortions, redshifts, but mistook their cause as velocity leading to the concept of an expanding universe—the Big Bang. A re-examination of these changes using a higher-dimensional curve leads to a more coherent theory. It also helps to explain the instantaneous speed of gravity, the measurable, but unreachable speed of light, a unification of all force, and the weird concepts of quantum theory ("spooky" action or entanglement, and wave-particle duality).*

*Electricity, magnetism, and the nature of electromagnetic waves are briefly discussed, as are other concepts—black holes, holograms, the large-number hypothesis, and virtual particles. Always the underlying thesis is that of a universal force—God.*

## Civilization—Efficiency And Overabundance

*Question:* You have put a lot of concepts together. At times they make sense; at times they seem muddled. Perhaps you could summarize your thoughts a bit and make them a little clearer.

*Answer:* I guess this is as good a time as any. Let me try. We started, because of my medical background, with a broad discussion of overabundance leading to heart and vascular disorders. We considered the basis of the problem to be society's ever-increasing efficiency, first found at the dawn of civilization, when humans began planting crops and herding animals. This, in turn, led to the dilemma of extra time (that not needed for day-to-day survival). Surrogates for this unnecessary leisure were required; games, drugs, excessive work, and religion have all been means of addressing this concern.

The most interesting and intriguing of all these useful diversions, to me at least, is religion. We alluded to a force that permeates all things, that allows for an accumulative process—in humans, felt as a constant hunger for more. This force appears to be within all that exists; those spiritually inclined often equate it with God. We then attempted to show how science has grown and has tried to further comprehend this force.

We went from Einstein's theory of relativity—space and time distortions as velocity increases—to a fourth-dimensional construct. We tried to show how a higher dimension is perceived in a lower one with changes in distance measurement. Then we discussed expanding redshifts and Hubble's interpretation as a velocity increase dependent on the distance observed.

*Question:* Let me interrupt. Once you started to discuss Hubble's findings of increasing redshifts with distance, you then proposed a radical change in accepted cosmology. You discarded the Big Bang theory and tried to explain it all as a fourth-dimensional curve. Since this is so different from current theory, perhaps you could refine it a little now.

## Clarification Of Hubble's Findings

*Answer:* Okay, let me try to explain my concepts. Hubble, and other astronomers, found ever-increasing redshifts the farther they peered into space. They based the redshift on a Doppler effect and, thus, concluded that velocity kept increasing the farther out one looked. Hubble perceptively graphed velocity against distance (up to about 100 million light-years) and found a relatively straight line—a continual increase in velocity the farther one gazed.

At this juncture, other scientists inferred that, an expansion of the universe was the cause of Hubble's findings and, therefore, if expansion was occurring, at some time in the past there was but a single point, a primeval atom that burst out and became our universe: the Big Bang. This theory was vigorously debated; however, in 1964 cosmic microwave background radiation (CMBR) was discovered and, as it so neatly fit the Big Bang concept, this conjecture became the accepted theory of the universe.

Other basic concepts also fit: the hydrogen/helium (H/He) ratio, and the immaturity of distant (early) galaxies. However, some disquieting problems arose. First was the uniform nature of the universe; essentially, it looked the same in all directions. A Big Bang would create a different picture, not nearly as smooth or homogeneous. To allow for this apparent conformity, very early and extremely rapid inflation was postulated. No reason could be given for this sudden, tremendous increase in volume; however, as it complemented the scenario so well, it became accepted gospel.

In the late 1990s, astronomers looking for a possible contraction of the universe used a *standard candle* (a special type of supernova—extremely bright and always of the same luminosity) to establish distance. What they found was the opposite, an expansion; objects were farther away than assumed by then current theory. To accommodate this newly found enlargement (beyond that even postulated by the Big Bang), a new force of energy was proposed—unknown, or dark, energy. However, so much dark energy was required that it alone accounted for about fifteen times as much energy–matter as the observable universe contained.

Thus, today we have a theory of the universe (the Big Bang), but for it to be meaningful we require unexplainable, early, rapid inflation, and great amounts of unknown or dark energy.

*Question:* Okay, let me restate what you are saying. Hubble saw continuously increasing redshifts with greater and greater distances and, therefore, assumed that objects were moving from us, and each other, ever faster the farther apart they were. This was believed to show an expanding universe and, thus, at one time in the distant past it must have all started as a single point. Weird, early inflation and unknown, repulsive energy were needed to explain the way it looks to current astronomers. However, didn't you leave out dark matter, the supposed substance that allows for the observed rapid rotation of galaxies; it alone is felt to contain almost five times the mass seen in the rest of the known universe?

*Answer:* You are correct. Besides dark energy, there is so-called dark matter, and the two are thought to make up over 95 percent of everything that exists. The problem is that no one has any idea what these dark entities consist of, and, thus, today less than 5 percent of the universe is real, or explainable, and over 95 percent is, in effect, fantasy.

*Question:* So if 95 percent is make-believe, why should one rely on the Big Bang in the first place?

*Answer:* I agree; it essentially makes no sense. The basic assumption by Hubble and other astronomers was incorrect; the redshift does not imply an increase in velocity, it describes a bend toward a higher dimension. When the math is done (simple tangents) the $z$ parameters (percentage increases in the wavelengths from distant sources of light) concur with this fourth-dimensional bend, and one can dispense with dark energy. Since there is no expansion, early inflation no longer need be considered and is also neatly discarded.

*Question:* If the Big Bang theory is incorrect and expansion is a chimera, how did you explain CMBR, the H/He ratio, distant (early) immature galaxies, and dark (unknown) matter? These were all part of the theory and seemed to fit nicely.

## Fourth-Dimensional Rotating Spheres

*Answer:* We used the concept of a rotating surface on a fourth-dimensional sphere. As three-dimensional entities, we see almost (barely less than) $90^0$ of this sphere. Just as it disappears, the stretch is so pronounced ($z$ is so huge—over $10^{32}$) that light waves are greater than the universe, and it no longer is visible. The last perceptible interval is the Planck length ($1.6 \times 10^{-35}$ m). If we were to expand this distance by the $z$ at this point, we go from $10^{-35}$ to about $10^{-3}$ meters (or microwave radiation). Thus, we explained CMBR as the lengthening at the edge of the universe—it is all-encompassing and constant.

We could not give a firm explanation for the H/He ratio, but that assumed by current theory (based on a nonexistent Big Bang and, subsequently, unnecessary inflation) is also suspect. The ratio still presents difficulties (specifically, a lack of lithium and an excess of deuterium) that have never adequately been explained. Therefore, given the universe's indeterminate age, it may just represent a relationship that occurs provided there is indefinite time.

Immaturity of early (distant) galaxies is extremely difficult to prove—they are very far away and barely perceptible. Quasar light sources, gravity lensing, and the excessive brightness of immaturity all call this model into question. Finally, dark matter, an idea that neither proves nor disproves the universe's expansion, is explained as the photonic mass ejected from the energy vortices about black holes.

*Question:* Using your concept of fourth-dimensional rotating spheres, you tried also to show the universality of all forces. However, you came up against a couple of really significant problems—the great disparity in their relative strengths, and the constant velocity of light versus instantaneous action. Why don't you now quickly summarize your thoughts?

## Universal Force And Velocity

*Answer:* Again, remember, I assumed, as did Newton, that gravity is instantaneous (permeating the entire distance of the universe—1.3 x $10^{26}$ m—in the shortest time possible, a Planck instant, 5.4 x $10^{-44}$ s, or 2.4 x $10^{69}$ m/s). But light speed is constant and measurable (3 x $10^{8}$ m/s). To equate them, the factor 8 x $10^{60}$ (Planck instants since the *start* of the universe, or Planck lengths in measuring its size) must be employed.

Since we perceive the universe to be three-dimensional and constantly moving forward in time, during each moment a complete world must exist. Thus, the force that holds it together, gravity, must re-establish the entirety each instant; it must be *instantaneous*. The force, however, as felt between instants travels through the higher plane (considered time—we cannot conceive of a fourth direction), and, therefore, it has traveled through 8 x $10^{60}$ separate moments. Thus, the velocity of this force (electromagnetism) is diluted by that number (from 2.4 x $10^{69}$ to 3 x $10^{8}$ m/s).

As to all forces being comparable, distance into the fourth dimension leads to differences in size. We cannot sense the distance, but we can see the change in extent. Protons are approximately $10^{-15}$ m, our universe $10^{26}$ m; the difference ($10^{41}$) is the ratio of strength between the strong force (holding protons together) and gravity (holding the universe together). Thus, we have evidence of equivalence between the strong interaction and gravity.

In a like manner, electromagnetic force, although appearing $10^{39}$ times greater than gravity, is really the same only felt through the fourth dimension. The *Fine Structure Constant,* an important but poorly understood ratio (1/137 or roughly $10^{-2}$), has, as one of its meanings, the difference in strength between electromagnetism and the strong force. Thus, electromagnetism has about one one-hundredth ($10^{39}/10^{41} = 10^{-2}$) the intensity of the strong interaction, but both are manifestations of (or the same as) gravity, only present in the fourth, not the third dimension.

*Question:* So your answers are tied to the fourth dimension. Velocity and size are constructs of this higher dimension. You also alluded to quantum weirdness in your discussions.

## Quantum Weirdness And Electromagnetism

*Answer:* Well, *spooky* (immediate) action fit my concept of instantaneous velocity within three dimensions, wave–particle duality made sense as a visualization of the phases of higher-dimensional objects as they rotate within our plane, and the two-slit experiment was explainable with force not particles.

Electricity and magnetism are how, in our world, fourth-dimensional spheres manifest (as concentric, rotating shells ever drawn to a central core), and the nomenclature of "positive" and "negative" is a misconception based on Benjamin Franklin's notion of electricity as a flow. Electromagnetic waves, although composed of photons, are actually the energy moving *through* these photons, and these particles have real substance, real mass. When photons strike electrons, they cause dislodgement (Einstein's photoelectric effect); they appear to be the answer to the riddle of dark matter.

I try to show that a proton represents a universe similar, in all respects, to ours. It too is a fourth-dimensional sphere with a rotating surface and, as it is visualized in our three-dimensional universe, it has about it ever-enlarging spheres of matter–energy. It is part of us, but we are part of it.

## Large Numbers And Virtual Particles

*Question:* Do you recall your thoughts concerning the large number hypothesis (ratios of $10^{40}$, $10^{80}$, and $10^{120}$) and virtual entities; why not review them now?

*Answer:* Let me try: $10^{40}$ is about the same as $10^{41}$, the ratio of the size of the universe to a proton. Thus, if we conceive of the universe as the surface of a fourth-dimensional sphere, that entity's radius would be roughly $10^{41}$ times, its surface $10^{82}$ times, and its volume $10^{123}$ times as great as a proton.

Therefore, if each proton represents (or is a portal to) a universe, there are close to $10^{82}$ such *real* universes that make up our three-dimensional covering (or world) but around $10^{123}$ *latent* universes that could exist. If each is our equal, the vacuum energy as described by quantum mechanics is the sum of them all—about $10^{122}$ times as great as that observable (I am assuming $10^{122}$ to be equivalent to $10^{123}$ for this example).

Also remember, the virtual particles with their antiparticles that fleetingly enter and exit our universe, leaving with a burst of energy, are best described as fourth-dimensional spheres that we see first on their outer surface then on their inner aspect. Since we cannot conceive of this higher-dimensional perspective, we view them as separate and opposite particles (inside-out, so to speak). Their burst of energy upon departure ($mc^2$) is but the reformation of the inertial cloud of substance; it is the counterforce of gravity with all its force (all the energy that is available in any particle).

Furthermore, when we mentioned the apparent scarcity of antiparticles, we ascribed it to the same concept. Since they are the inverse (the inside out) of particles to see them we must obtain a fourth-dimensional viewpoint. This only occurs when a virtual particle enters the third dimension or when an actual particle is pushed toward the fourth. Since these occurrences are at best fleeting, normally all we find in our three-dimensional world are real particles—antiparticles are quite rare indeed.

We also talked about the illusion of motion; how fourth-dimensional objects leave a footprint on our three-dimensional surface. The surface itself is what travels at the speed of light; the closer an entity is, the faster its movement and the more fourth-dimensional its appearance (smaller in length and denser in mass).

## Black Holes And Holographic Surfaces

*Question:* Since you are describing our world as a surface phenomenon, why not also review your thoughts concerning surfaces and the transfer of information?

*Answer:* Okay, let's make an attempt. When we discussed the concept of a black hole, we noted that the total information of such an object (the electromagnetic radiation describing all its physical qualities) is found on one-quarter of its surface or event horizon. If a black hole describes our universe (a fourth-dimensional sphere), we are always that one-quarter of its exterior; we, therefore, contain all the information that exists in its entirety.

Finally, if we view the universe as a hologram, each part has all the information of the whole. The smallest part has Planck-length dimensions, and there are $10^{122}$ such parts. This equates with the total of real and virtual protons (universes), and, therefore, each has all the information of the complete entity—all are entangled.

*Question:* Allow me to summarize your thoughts. You appear to be saying that, as we gaze ever farther out, we note a higher-dimensional curve; thus, we can infer a fourth spatial direction to our world, allowing for a universality of force, a constant but unreachable speed of light, and an instantaneous velocity of gravity. The same force that forms and holds all things together also accounts for accumulation and growth—it allows for mankind's ascent and attempt to understand. Once perceived, it becomes the basis for our contemplation of God.

*Answer:* Your interpretation is concise but reasonable. Let me now attempt to explain how I view the current theoretical landscape.

# 33

## BASIC MISCONCEPTIONS

I N THIS CHAPTER WE RESTATE our fundamental concepts and show how much of what we see about us then makes more sense. Our world is the surface of a 4D sphere traveling at the speed of light, c. All force is based on the centripetal effect of this motion.

Universality of force stipulates that each proton represents an individual world just like ours. Each contains all of the others; thus, the total quantum vacuum energy is approximately $10^{120}$ times as great as measureable—it is the energy of all protons (all universes) both real and virtual.

Given these concepts, supposed explanations concerning the increasing redshift (CMBR and dark energy) are now explainable. One gets a clearer picture of electricity, magnetism, and light; magnetic monopoles disappear, along with a need for baryogenesis (the magical loss of antimatter). Other problems, including quantum weirdness (spooky action and wave–particle duality) can be understood. Allowing photons to have mass sheds "light" on dark matter and helps explain galactic rotation.

Much of particle physics (the Standard Model) is shown to be based on a misconception—an expanding, hence cooling, world. Its explanations and attempts at discoveries (proton decay, magnetic monopoles, Higgs boson) are, therefore, fruitless endeavors—ultimately destined to end only in frustration.

## Our Three-Dimensional Surface World

Let me try now to restate my ideas in order to give a clearer picture to things as I see them. The universe is the three-dimensional covering of a fourth-dimensional sphere. It can encompass just under one-quarter (almost $90^0$) of that exterior. Less than $0^0$ would connote the unknowable future; $90^0$ or more, and all things would disappear. This surface moves toward the fourth dimension, toward time, at a constant rate—$c$ (the speed of light, $3 \times 10^8$ m/s). This motion defines centripetal force, a pull toward an imaginary or unknowable center. We understand this force as gravity.

Each new instant brings a different, somewhat unique world. It lasts the shortest time possible—a Planck moment ($5.4 \times 10^{-44}$ s). Since the universe reforms each time, gravity, the force that re-establishes the world, must travel its entire distance ($1.3 \times 10^{26}$ m) each moment; therefore, its velocity appears, in essence, to be instantaneous—or $2.4 \times 10^{69}$ m/s ($1.3 \times 10^{26}$ m / $5.4 \times 10^{-44}$ s).

Electromagnetic force, on the other hand, travels from moment to moment over time. As there are $8 \times 10^{60}$ such intervals making up the totality of our universe, electromagnetic force is diluted by that factor and appears to be traveling at a measurable but unreachable velocity—the speed of light ($3 \times 10^8$ m/s = $2.4 \times 10^{69}$ m/s / $8 \times 10^{60}$). Both forces, however, are, in reality, the same, manifestations of the centripetal pull established as our world rotates through time, and both represent (or cause) the entire energy of our world. Thus, if:

$$\text{(Centripetal force) } F = mv^2/r, \text{ and if}$$

$m$ is the mass of our universe, and if
$v$ is its velocity $c$, and if
$r$ is an imaginary or 4D direction ($i$), then

$$F = mc^2/(i), \text{ or}$$
$$(i)F = mc^2, \text{ and since}$$
$$E = mc^2, \text{ then}$$
$$(i)F = E.$$

Therefore, the energy of the universe ($mc^2$) equates with (or, is a result of) the inwardly directed force seen as gravity within the third dimension, and as electromagnetic force through the fourth.

## Unity Of Force

*Question:* Okay, you have restated your basic ideas. What about the fact that gravity is much, much weaker than the electromagnetic force?

*Answer:* Since the universe has four spatial dimensions, but we can conceive of only three, the fourth direction, if understood as a distance, would have to be the smallest possible—a Planck length (or $1.6 \times 10^{-35}$ m). Thus, things far away in that spatial dimension appear within what already exists, but much, much smaller; whole other worlds become mere protons. As our universe is approximately $10^{26}$ m, but a proton (when presenting three-dimensionally) is about $10^{-15}$ m, the difference ($10^{41}$) is felt as the disparity in the force holding realms together—the centripetal pull.

Therefore, the strong interaction (the force that holds hypothetical quarks within protons or neutrons, or those nucleons within larger nuclei) is really the same manifestation of centripetal force seen in gravity, only appearing much greater since it is so exceptionally concentrated. It is gravitational force in a distant (tiny) fourth dimension. However, just as each proton represents the portal to a separate universe encased within our world, so too is our world found within each proton. All are entangled; all have knowledge of all others.

## The Meaning Of Large Numbers

The universe is $10^{41}$ times as large as a proton; therefore, it takes $10^{41}$ protons, end to end, to equal its diameter. Since we are the surface of a fourth-dimensional sphere, there would be about $10^{41}$ squared (or $10^{82}$) such entities within that surface. Finally, the volume of that sphere would contain approximately $10^{41}$ cubed (or $10^{123}$) separate protons, of which most ($10^{41}$ or 100 thousand trillion, trillion, trillion times as many) would be within the fourth dimension, or beyond our tangible space. However, as each represents a separate universe, with the same energy as ours, the total energy of the virtual world (the theoretical quantum energy of the vacuum) would be about $10^{123}$ times as much as our observable world.

**Question:** So what you are saying, thus far, is that the fourth dimension can only be conceived of as time; we cannot understand it as a physical direction similar to the other three. However, it exists spatially; therefore, other universes appear much smaller but within our world. The number of protons (estimated at $10^{80}$) equates, more or less, with the surface of a great fourth-dimensional sphere, our knowable universe, but the total quantum energy equates with the volume of that same sphere.

## Redshift And The Big Bang

**Answer:** You do clear up my concepts somewhat. But let me continue. Since Hubble's discovery of increasing redshifts (with distance), the usual assumption has been that the universe is expanding. This led to the concept of a Big Bang—everything derived from the explosion of some primeval atom. The theory was accepted by many when cosmic microwave background radiation (CMBR) was found and explained as due to the continuous stretching of light from the initial event to the present. It was additionally strengthened—with the explanation of the hydrogen/helium ratio (H/He~75/25)—as what individual protons and neutrons would do as the universe expanded and cooled, and further enforced by the supposed evidence of immaturity in distant or early galaxies.

However, although these pillars form the underpinnings of the Big Bang theory, cracks keep forming in its foundation. Expansion cannot account for the lack of theoretical magnetic monopoles or for the sameness, the homogeneity, of the universe. To solve these problems, a theory of very early, extremely rapid inflation was patched onto the expanding universe. More recently, when looking for a probable contraction of our universe, astronomers using a standard candle (a specific type of very bright supernova) to determine distance found, instead, an unexpected greater expansion. To cover this awkward discovery, dark energy was introduced as a support.

*Question:* You have discussed all this before. What are you now trying to show by bringing it back again?

## Redshift And The Higher Dimension

*Answer:* You are correct; but by using my initial assumptions, I will attempt to give a better or clearer picture of what is about us. If we go back to my original concept, our world as the three-dimensional surface of a fourth-dimensional globe, then the above findings make more sense.

First, there is no expansion. The redshift is simply how the higher dimension is perceived—increasing lengths the farther out one looks. Thus, if there is no expansion, there is no need for the Big Bang theory, or any other theory, to explain what does not exist. CMBR is more easily understood as the stretch of our visible universe at its farthest edge. Inflation is obviously not required, as there was no event in the first place. Finally, dark energy, that mysterious force of even greater expansion, disappears; it is but an illusion caused by a misjudgment of distance using the Hubble law as a proof of expansion.

*Question:* Again you leave out distant immature galaxies and that annoying H/He ratio. These are hard to refute and appear to be significant bases for the Big Bang concept.

*Answer:* I put them in grudgingly, as I have no concrete way of refuting them. However, they are not as solid as one might think. Remember, as astronomers use better and more sophisticated methods of deep-space exploration, unexpected maturity in ancient galaxies keeps popping up. The *James Webb Space Telescope*, scheduled for launch about 2018, should be able to resolve many of these distant entities and I strongly feel will show complete development where, theoretically, none should exist.

The H/He ratio, although apparently explainable by a cooling or ex-panding early universe, still has problems with the dearth of lithium and the abundance of deuterium. So even that pillar of the Big Bang theory is not without flaws.

## A Clearer Picture

However, if we assume my scenario to be correct—that we are, in essence, the three-dimensional surface of a fourth-dimensional globe—then many other things can be understood. More mundane concepts such as elec-tricity and magnetism come into clearer focus. One can do away with the silliness of magnetic monopoles. Magnetic attraction is easy to visualize. The concept of "positive" and "negative" can be discarded as a relic from Benjamin Franklin's era. Virtual particles make sense. The paucity of anti-matter is easy to explain, and there is no longer a need for baryogenesis (that mysterious destruction of early opposing entities magically leaving just enough of current day's material but very little of its opposite, antimatter).

In a similar manner, quantum weirdness need not be so strange. Spooky action at a distance can be grasped, as can wave–particle duality. The two-slit experiment becomes understandable. The nature of light can be clarified by allowing photons to have mass, as all real particles in our world must. Even dark matter has *light* shown upon it if we under-stand it as that same photonic *sea*.

The so-called cuspy-halo problem (why dark matter should theoreti-cally be at the center of a galaxy but instead appears at its periphery) can be explained with the same massive photonic encirclements found about any black hole. The shape of a galaxy is also better understood as due to the ejecta of material swirling, in a similar fashion, about a central core.

## Information

*Question:* So if I understand what you are saying, our universe, al-though appearing to have but three spatial dimensions, really has a fourth. It is the surface of a sphere in that dimension. To you this sphere is a black hole and our world is its event horizon. You showed earlier that theorists claim one-quarter of the exterior of a black hole contains all of its information. Since we cannot be more or less than that one-quarter, we thus contain all the information of the totality.

*Answer:* As always, you are quite astute. That is my basic concept. All the information of that great sphere, of which our knowable universe comprises but a partial cover, is contained within our world. All worlds are within all others; all have the entire information of the whole.

## The Standard Model

*Question:* So if one agrees with your ideas, much of today's astronomy and physics needs to be discarded. Existing models appear as Ptolemaic schemata when first pierced by Copernicus.

*Answer:* Yes, much of current theory is founded on a basic flaw; one that, as it is so fundamental, is so hard to uncover. The Big Bang would make sense if there was an expansion; as there is none, it is superfluous. A lot of theoretical physics is based on that same, unnecessary assumption. Their notions, thus, become illusory. Cracks keep opening in these concepts; however, they continue to be closed with newer, more imaginative (or inane) solutions.

The Standard Model, the currently accepted theory of particle physics, is constructed on the belief of an expanding, hence cooling universe. In the primordial cauldron, photons, matter, and antimatter (all types of energetic and exotic entities) are fully mixed together. As the temperature falls, as the world enlarges, different particles condense out.

The concept is elegant and logical but, just as was Ptolemaic theory, its basic assumption is flawed. There was no Big Bang; thus, there has been no expansion and cooling. Particles do not congeal out from some primeval ooze; they exist as fourth-dimensional objects. Matter and antimatter are shown to be two aspects of the same entity simply viewed in our world from differing perspectives.

The basic rationale, CMBR, is more easily explained as the stretch at the edge of our world. The supposed photon/proton proportion of one billion to one (used to explain the basis of H/He and other elemental ratios), then, shows itself to be nothing but an *ad hoc* number, patched onto a theory, that allows for accordance with nature.

*Question:* But what about the recent discovery of the Higgs boson, the particle that is thought to give mass to all that exists—the *God particle?*

*Answer:* I am glad that you bring it up. The whole search for a Higgs boson (supposedly just discovered at a cost of many billions of dollars) is a pointless endeavor. It, like much of modern theory, flows from the concept of an expanding, cooling universe. It, like the magical magnetic monopole, is a chimera—a quixotic search to prove a theory that is based on a nonexistent foundation.

The supposed limited life-span of protons is also a misunderstanding; that is why their decays are never found. Vast sums have been fruitlessly spent looking for their demise. However, they are fundamental constructs of other universes and most likely are, just as we, eternal.

*Question:* So if you do away with an expanding universe, you remove, not only the Big Bang, but also all the physics that explains the nature (the formation and maturation) of particles in an ever-enlarging and cooling world.

*Answer:* You are correct; the Big Bang theory and much of modern physics is founded on a basic misconception. Let me explain this a little more in the following chapters.

# 34

# MODERN MYTH

J UST AS BIBLICAL LEGEND CONTAINS a beginning and subsequent fall, modern myth has the Big Bang followed by a breaking of perfect symmetry. Although the logic associated with current theory is elegant, the assumption of a Big Bang is faulty. As there is no initial explosion leading to expansion and cooling, all of particle theory based on it descends into story—or myth.

## Expansion And Cooling

Let me more carefully go through modern physics, and its inherent mythology, and see if our explanations make more sense. The Standard Model presupposes an initial Big Bang. As the universe subsequently expanded, it cooled from an unbelievably high temperature to what it is today. This change allowed for different particles and forces to condense and separate.

*Question:* I see, like water forming on a cold soda can during a hot summer evening.

*Answer:* Yes, that is quite similar to what is described. The higher the temperature is, the more dynamic the particles. On a hot, humid evening, water vapor remains energized; it stays in the air. Only when contacting a much cooler surface—our soda can—do the particles lose energy and liquefy.

In current theory, at exceedingly high temperatures, energetic photons continuously collide with one another, making for matter/antimatter pairs that then shatter, forming new photons. At these temperatures, the universe is just a bizarre soup of dynamic ingredients. As expansion continues and temperatures fall, heavier, more massive entities (protons and neutrons) initially form. There remains ever less energy in the enlarging, or lower-frequency, photons to allow for further, ongoing breakage of particles into constituent parts (quarks and gluons). Ultimately, as the temperature persists in dropping, lighter, or less massive, objects (electrons and neutrinos) finally appear.

Supposedly, an exceedingly energetic, hot mixture, a perfectly symmetrical mélange, was broken apart due to expansion and cooling. What we are then left with is our current world.

## Biblical Background

*Question:* You know, the more I think of this, the more it resembles the Biblical narrative of Genesis. The Big Bang is creation, and the breaking of perfect symmetry our expulsion from the Garden of Eden.

*Answer:* Yes, the mythology of today's science is not that different from our religious beliefs. Initial perfection has devolved to our present state.

Now, the reason this story is accepted by most scientists is that it is based on reproducible experiments. If photons are of high enough frequency (energetic enough), their collisions can lead to matter/antimatter particles, which then break and reassemble as photons. Since this is seen at energy levels currently attainable, researchers feel that, by just pushing these values ever higher, all that is discussed should be conceivable.

It is a very elegant and logical belief; however, so too was Ptolemaic theory. Perfect spheres in the heavens allow for beautiful constructs but do not make for reality. Thus, just because something seems so sensible does not mean that it is correct.

*Question:* Okay, current particle science to you is mythology; it is based on creation—a Big Bang, and then a fall or breaking of perfect symmetry. So, if you consider it mistaken, how do you replace it?

## The Legend Of Symmetry

*Answer:* We have discussed the concept of a higher dimension, that our world is but the exterior of a sphere in that realm. All energy is secondary to the rotation or centripetal pull on that surface. All forces, then, are different manifestations of that rotation.

Protons represent doorways to distant (higher-dimensional) *mini-universes*, similar to our own. The supposed quarks and gluons (subnuclear material thought to make up protons and neutrons), therefore, could be conceived of as fourth-dimensional entities, constituents of these nucleons. These objects only theoretically exist (they have never been found), but the perfect symmetry of current theory calls for their presence.

Of course, just as quarks with gluons are thought to be real, the proton is also thought to decay. But a universe cannot disappear; it represents all that exists, and each world encompasses all others. The loss of one means the loss of all. Thus, the components of protons and protons' theoretical destruction are both part of a refined, elegant logic; however, the underlying concept is false. The premise, the Big Bang, is erroneous, and just as astronomers look for dark energy, so too is the search for decay and constituent parts pointless.

Current theory predicts magnetic monopoles; the math demands their existence; yet they have never been found. Again, like a proton's demise, magnetic monopoles presuppose the breakup of an entire universe, something that cannot happen.

The mathematics associated with perfect symmetry also calls for a Higgs mechanism (something that gives mass to particles as they condense out of the primordial cauldron). But there *is* no primeval soup; there is no explosion or expansion, no cooling or condensing of matter from some perceived perfection.

So the searches undertaken for the Higgs field, with the supposed recent discovery of the Higgs boson, are similar to the quests for the existence of magnetic monopoles or proton extinction—pointless endeavors. Much has been, and will be, spent; many jobs are dependent on this enterprise; but to no avail. It is a search for a chimeric beast and will end only in irrelevance and frustration.

*Question:* So if there was no Big Bang, no expansion or cooling, then there was no perfect symmetry to break—no proton decay, magnetic monopoles or Higgs mechanism. Much, then, of our pursuit is fruitless.

## Mathematical Delusions

*Answer:* Yes, science has gone off on a tangent leading nowhere. It is based on ever more complicated mathematical concepts that are logical and rational, but that rest on the fundamental error of an expanding universe. The Big Bang was, after all, formulated to explain the redshift increase seen with distance. We have shown this theory to be erroneous. Thus, all that follows must also be invalid.

Inflation, a wholly *ad hoc* addition to make the Big Bang acceptable, has been shown by one of its earliest adherents, Paul Steinhardt, to be, essentially, infinitely impossible. It goes against the second law of thermodynamics—of ever-increasing entropy or disarray. It is even less probable than the aforementioned cup of coffee falling off a table, breaking to bits, spilling its contents, then miraculously reforming with liquid intact on the same surface.

Inflation was patched on to explain the difficulty in finding magnetic monopoles, creatures that should have existed yet could not be located. It was also needed to explain the sameness—the flatness and homogeneity—of the universe. But as there was no Big Bang, there is no need for this pointless add-on.

## Photon/Proton—Hydrogen/Helium

*Question:* Cosmic microwave background radiation (CMBR) is always brought up to give credence to the Big Bang. It is probably the most important foundation, after the redshift, for this theory. Nevertheless, you have shown it is simply how the stretch at the edge of our world is perceived, how the last distance possible, the Planck length, is expanded to microwave length. But wasn't it the CMBR that gave rise to the current explanation of the hydrogen/helium (H/He) ratio? Doesn't it have to do with the number of photons and protons in some way?

*Answer:* You are astute as always. The ratio of photons to protons is theoretically the cause of hydrogen fusing to helium. Photons carry energy, and if there are enough of them at a high enough temperature, fusion can occur. This ratio is assumed to be about one billion to one.

However, this proportion has not been exactly determined; the extent of protons can vary by a factor of two to three. Thus, no one is quite sure of their real amount although it is thought that, because of CMBR, the quantity of photons is known. But, as already noted, CMBR has nothing to do with a Big Bang; the initial or photonic energy of the world has not been stretched out to today's size. Photons, I feel, are not even what most scientists think they are—they have mass and do not move at *c*. They are the ejecta from energetic whirlpools surrounding portals to individual worlds or protons.

As there was no Big Bang, no expansion, no inflation, and as CMBR is misunderstood, so too is the cause of the H/He ratio. It is not due to some extrapolated photon-to-proton relationship (which never really gives the correct lithium ratio); its cause is to be found in the eons of time our universe has existed. I am not capable of venturing more than a guess, but I know current theory is incorrect.

## Distance And Immaturity

*Question:* Another foundational support for the Big Bang theory is the supposed immaturity of distant galaxies. But you have shown that too is a shaky pillar.

*Answer:* Yes, although it was presumed that distant objects, entities supposedly formed closer to that primeval event, were less mature, as astronomers utilize more powerful and sophisticated means of observation a clearer picture of that earlier epoch emerges. Using distant quasars as light sources and intervening galaxy clusters for extra magnification, maturity—metallicity, or the presence of complex elements—is seen where none should theoretically exist. What was originally discovered, evidence of bright, young-appearing galaxies, is now seen to be adulterated with duller, less intense, more established objects. Simply seeing better allows one to see dimmer (or older) entities.

As stated previously, observation will trump theory. The *James Webb Space Telescope*, scheduled for launch about 2018, will open up our universe as did the powerful Hubble apparatus before. I am sure that, once this telescope is finally in use, the universe will appear quite unlike what is currently envisaged. What we will see in the far distance will be similar to what we now see nearby; the Big Bang will be forgotten, as was Ptolemaic theory and ancient Greek science.

## Rebirth Of Science

*Question:* So if these pillars disappear—redshifts signifying expansion, the misinterpretation of CMBR, the almost infinitely impossible concept of inflation, the inaccurate basis of the H/He ratio, and the incorrect assumption, in distant galaxies, of immaturity—then there was no Big Bang. Its lack, therefore, changes particle theory, does it not?

*Answer:* Yes, current theory with its symmetrical Garden of Eden will be seen as myth once Big Bang genesis is gone. There will be no primordial soup of condensing particles. There will be no miraculous antimatter annihilation or baryogenesis (a process that magically destroys exactly enough unwanted material, leaving just normal matter to fill the void). We have already shown how a particle and its antiparticle are the same when viewed from a higher-dimensional perspective. We have already explained why, then, matter is all that commonly exists; antimatter is rare and can only be found when violent events push particles from our world (the three-dimensional film that covers a fourth-dimensional globe).

Current theories in astronomy and physics, therefore, are about to topple. They have been erected on a fundamental misunderstanding. Science has taken a long detour of close to 100 years but will soon regain its footing; there will be a rebirth.

# 35

# ILLUSORY PROBLEMS

B*ASIC CONCERNS ASSOCIATED WITH THE BIG BANG THEORY include the flatness, horizon, structure, and relic problems. Theorists employ the concept of inflation as a solution. We discuss the essential impossibility of this scenario and describe how, with the use of a higher dimension, these problems are shown to be illusory and disappear.*

*Grand Unification Theories also are briefly discussed and shown, as was the Standard Model, to be elegant, logical concepts, but, being based on a mistaken premise, invalid.*

## Concerns

**Question:** In the last chapter we discussed problems with particle theory and how basing it on the Big Bang leads to fundamental mistakes. Could you clear up some other concerns I have come across in my readings—specifically, the flatness, horizon, structure, and relic problems associated with the Big Bang?

**Answer:** We have gone over some of these already, but perhaps I have not been as clear as possible. Let me start with your first concern.

## Flatness Problem

If the universe began in a Big Bang, then, according to Einstein's theories, it most likely would be contracting or expanding at the present. These scenarios are dependent on the matter–energy content of our world today. If that amount, when compared to a critical density of gravity, were too great, it would cause contraction; if too small, expansion.

However, precise measurements with current space probes lead to a third possibility: one that is neither—one that is flat. These accurate findings show our world to be within one percent of the density needed, if there had been a Big Bang, to exactly counter the contraction of gravity.

The reason it raises a problem with the original Big Bang theory is that, for such a present close approximation, in the distant past, just after the supposed primal event, the closeness had to be much, much more exact. Instead of 1 percent or 1 to 100, it would need to have been 1 to $10^{62}$ or 1 to 100 trillion, trillion, trillion, trillion, trillion, odds so remote as to be, for all practical purposes, impossible.

**Question:** So the flatness problem is based on the concern that, if the universe began in a Big Bang, it should not be flat now.

**Answer:** Yes, that is the gist of the problem. Given a Big Bang and Einstein's theories, the probability of a flat universe is vanishingly small. Now, the actual matter known to exist is less than 5 percent of the amount needed, if there had been a Big Bang, to flatten our world. Dark or unknown matter and energy have been postulated to make up the extra 95 percent, bringing the universe to its current flat existence—the sweet spot between expansion and contraction.

If, at primordial time, this balance was off by even the slightest amount, we would not exist; our current world would have either already contracted (back to an infinitesimal point) or expanded (so rapidly as to not have formed at all). But we and it are here; thus, there has to be a solution.

*Question:* Since we exist, and since our world is measurably flat, how do scientists explain this extremely unlikely occurrence?

## Inflation

*Answer:* Current theory uses the concept of inflation—an extremely rapid, very early expansion. Since it was so swift and huge, any curvature in the early universe at the present would appear flattened. It is similar to a wrinkly balloon being blown up; once fully expanded, all the creases are gone. Therefore, our primitive world's matter–energy content did not have to be anywhere close to the critical density for flatness to now exist.

*Question:* But didn't you already state that even some of inflation's adherents claim the theory to be untenable, that it is essentially impossible?

*Answer:* Yes, we noted the article in *Scientific American* by Paul Steinhardt that claimed the odds of inflation having occurred were 1 to $10^{(googol)}$, a number so great as to be unwritable. However, that is not the real reason the flatness problem is foolish. There was no Big Bang in the first place; there is no expanding universe; therefore, there was never a time when it was very, very small and the ratio of density to critical mass needed to be so finely tuned. Thus, there never was a problem to begin with, and the solution—inflation—was never required.

## An Illusion

In today's world, what we consider flatness is just an illusion. The universe appears flat in three-dimensional terms; however, flatness is simply all we can envisage. To find the real bend toward the fourth dimension, we must visualize an imaginary direction, and we cannot. All that can be noted is the redshift, or the alteration caused by that higher dimension. Of course, if we consider that distortion to be a velocity increase (consistent with expansion), then the subsequent enlargement would be judged flat. We cannot sense a bend; hence, when present, it appears deformed or flattened.

Therefore, flatness stems from our inability to see a higher-dimensional curve; it is inherent to our design. The universe must appear flat, since we can only conceive of it that way.

*Question:* Are you saying, then, there is no flatness problem, that the real problem is with our inability to sense another dimension?

*Answer:* Yes, there was no Big Bang. The universe is, as already shown, the lower-dimensional surface of a higher-dimensional sphere. We look into the distance and notice an increasing redshift; we explain it in our own terms (as increasing velocity, as expansion). However, this is because all we can understand is our three dimensions; a bend toward a higher plane is unknowable.

Flatness is an illusion; it is caused by assuming the four-dimensional world to have but three. The more carefully we measure in our plane (the lower dimension) the flatter will be the result. It has nothing to do with matter–energy content versus a critical density; it has only to do with an inability to properly comprehend the redshift as a higher-dimensional bend.

## Horizon Problem

*Question:* What about the horizon problem? If I understand it, cosmic microwave background radiation (CMBR) looks the same in all directions. But energy in distant, widely separated locations (sites that never had time to intermingle) should not be identical.

*Answer:* You do seem to understand the problem. CMBR is essentially the same in all directions, over distances too great to have come into contact in the limited time since the Big Bang. From any viewpoint, CMBR is similar, even thirteen-plus billion light-years toward opposite poles. However, these poles are over twenty-seven billion light-years apart, and the universe's lifespan is less than fourteen billion years; therefore, light (or electromagnetic energy waves) could not have traveled from one to the other. Thus, there should be no reason for interaction (or equivalence). Why, then, are the energy levels the same?

Big Bang theorists answer this concern again with inflation. Areas in equilibrium instants after the primal event expanded outwardly so rapidly (many times faster than the speed of light) that initial equivalence was maintained and spread throughout. However, there are two fundamental problems with this solution. The first is a misunderstanding of CMBR; the second is the lack of any underlying expansion.

## Cosmic Microwave Background Radiation (CMBR)

CMBR is not due to the expansion of freed photons once atoms condensed from a hot plasma soup. It is only the *noise* of the very edge of our universe, the last Planck interval stretched to microwave proportions. Thus, it is, of course, the same in all directions. The universe has an edge just before it disappears into the fourth dimension, and this border surrounds all. It is only one Planck length in size, the smallest possible extent allowed. When stretched, it is perceived as microwave radiation, and, as it is the boundary, it is seen and is the same wherever noted.

Once we understand the real basis for CMBR, no silly inflation need be interposed. Thus, the horizon problem is due to a misunderstanding of CMBR; inflation need not apply. Of course, as with flatness, the essential reason for ridding ourselves of the horizon problem is the nonexistence of expansion in the first place. A problem needs no solution if it does not exist.

## Structure Problem

**Question:** So you have dispensed with the first two questions; what about the so-called structure problem?

**Answer:** The structure problem is similar to the first two conditions. The question it raises is, why is the universe uniform in structure throughout (why are galaxies as evenly distributed as they seemingly appear)?

According to the Big Bang theory, galaxies formed from inhomogeneous areas already present in the early CMBR, irregularities initially caused by the quantum fluctuations of inflation that had subsequently expanded exponentially. Galactic structures, therefore, during the inflationary event, were uniformly spread throughout the universe (individually and in clusters) and subsequently grew to their current status as the universe continued to expand.

We have already shown, however, that CMBR is not what the theory suggests. Galaxies did not form from these aberrations. Galaxies have existed for eons, as has the universe. Their formation is secondary to accretion of protons' black holes to much, much larger size.

Remember, each proton represents a universe encompassing all others, and each has been around for a very great time; thus, their agglomeration into galaxies is not to be found in meaningless CMBR fluctuations. Again, the whole problem is a nonexistent one, only present if the Big Bang is postulated. Since there was none, there is no problem.

## Relic Problem

*Question:* The last concern that is often raised is the relic problem. I think it has to do with unusual particles—magnetic monopoles—that would have formed as the universe expanded and cooled. They should be abundant, but there are none.

*Answer:* Yes, the relic problem has to do with phase changes during the expansion and cooling of the so-called Big Bang. However, because there was no such event and these phase changes—crystallizations—could not have occurred, then no magnetic monopoles need to exist. We have already shown that their very presence is impossible. It entails cutting a black hole in half—destroying a universe, and with it destroying all. If they existed, our world would not.

Of course, Big Bang theorists employ the ever-present inflation scenario to explain that, although these entities were abundant prior to inflation, they have since been lost (in the vastness of our universe). As with all other explanations, this one is also impossible. There was no Big Bang, there were no phase changes leading to relic particles, and there was no inflation scattering them throughout, making their discovery impossible. They have not been found since they do not exist.

*Question:* So the problems broached, and supposedly solved, in Big Bang theory are illusory. Neither the Big Bang nor concerns associated with it exist.

*Answer:* That, in a nutshell, is the solution of these so-called problems. They are solved because they do not exist in the real world.

## Baryogenesis

Other difficulties raised by using Big Bang concepts in particle physics are also likewise solved. Baryogenesis, the mysterious loss of antimatter, leaving our world filled with its opposite—matter—is one of these concerns. We have already shown that matter and antimatter are the same when viewed from a fourth-dimensional perspective. Our universe is a three-dimensional surface; all attached (stable) fourth-dimensional entities project onto it as three-dimensional. Only when that surface is pierced is the higher-dimensional particle seen first from the outside, then the inside. Only then do we see antimatter accompanying matter.

Virtual particles continually pop out of the fourth-dimensional abyss and just as readily re-enter after our observation of their outside and inside. We see a particle and its antiparticle, and we think they destroy one another. But they merely burst the surface of our world to disappear again, and the energy of the closure, as always, is equivalent to their mass $(mc^2)$.

## Grand Unification Theories (GUTs)

*Question:* So baryogenesis is similar to the other problems, a misunderstanding due to a basic misconception—the Big Bang theory. What about the so-called Grand Unification theories (GUTs), or theories of "everything"? Do they make any sense?

**Answer:** Like all the rest so far described, they have the same fatal flaw. GUTs postulate a breaking of symmetry as the universe expanded and cooled, leading to four fundamental forces. First, gravity was separated; then the strong force disconnected from electromagnetism. This supposedly occurred before inflation, quite early in the lifespan of our world. In fact, the breakage of forces is theorized to have led to this new inflationary stage via so-called *vacuum energy* causing hitherto unknown *repulsive gravitation*.

But we have shown that the universe has never expanded, and further that inflation is for all practical purposes impossible. Thus, the scenario explained by GUTs is meaningless. All forces are really the same (different manifestations of the centripetal pull); therefore, all must be attractive, and repulsion, then, would be impossible.

Gravity is how we describe force within the three-dimensional world; electromagnetism is the same but in the fourth dimension—between individual three-dimensional moments of existence. The strong force is simply gravity in a far distant fourth-dimensional universe (represented by a proton), and finally the weak force is only part of electromagnetism.

In fact, the original problem—flatness—is simply a misunderstanding of the universality of current forces. Since they are all from the centripetal pull ($mv^2/r$) total force equals $mc^2/(i)$ (where $v$ equals $c$ and $r$ is imaginary [$i$]). We then showed that, since gravity is equivalent to $(i)F$, it thereby equals $mc^2$ (all the energy of the universe). Thus, the force of gravity *must* equal the matter–energy content of our world; it *is* that entity. This is basically why, in three-dimensional terms, the matter–energy density equals the critical density of gravity. They are really the same.

Now, GUTs postulate proton decay and magnetic monopoles, but neither of these is possible—they simply cannot be. They both imply the destruction of an entire universe, and, as all worlds are entangled, as all are a part of each other, the loss of one would mean the loss of all. If protons decayed, or if magnetic monopoles existed, we would not.

*Question:* So everything discussed thus far is unreasonable, since everything is based on the supposition of a Big Bang. Without that occurrence all the rest falters.

## Hydrogen/Helium (H/He)

*Answer:* Yes, that is definitely the case. Let us look at the vaunted H/He ratio, probably after the redshift increases and CMBR the next most important pillar of current theory. This ratio (75/25) is based on the photon/proton proportion of around one billion to one ($10^9/1$). It is due to an estimate of total photon density unchanged since the primal explosion. Once plasma by cooling led to atoms and this radiation was freed, it, with the rest of the universe, expanded to its current size. Near-infrared ($1000 \times 10^{-9}$ or $10^{-6}$ m) photons grew by 1000-fold to microwaves ($10^{-3}$ m).

But expansion never happened, so the concept of CMBR apparently measuring these microwaves is incorrect. Therefore, the energy density (the photonic density) of the universe is misunderstood. Also, the actual number of protons is only an imprecise approximation (possibly off by a factor of two to three). So the supposedly established proportion of $10^9/1$ becomes somewhat illusory.

Thus, the very notion of the H/He ratio being an underpinning of the Big Bang appears faulty—it seems to entail a form of circular reasoning. The photon/proton proportion was simply an estimation used to get an H/He ratio that coincides with today's world; but, as CMBR's real cause is misinterpreted, the basis for this proportion's existence is also suspect.

## Distant Immature Galaxies

*Question:* The last of the major pillars of Big Bang theory, the apparent immaturity of distant (supposedly young) galaxies, has also been shown, has it not, to be quite shaky? Didn't we note advanced age among our so-called juvenile entities?

*Answer:* We certainly have, via quasar light sources and gravitational lensing. Again, once the *James Webb Space Telescope* is successfully launched (hopefully around 2018), distant objects will be much clearer. We will see a similarity of these entities to those close by. This important scientific advance will finally shatter the Big Bang theory and with it all the foolishness promulgated since its inception, including current particle and Grand Unification theories.

*Question:* You certainly have answered some of the questions I had, and I think you clarified your thoughts somewhat. However, if you are so set against current theories concerning the origin of matter and energy, why not try now to explain how you think things come into being—why they exist?

# SECTION IX

## *Reformulate*

**T**HE HYDROGEN ATOM, A 3D PROJECTION *of the essential component of the universe (the 4D hypersphere), is composed of individual particles (a proton and an electron) each representing different aspects of this higher-dimensional reality. Most particles (with the exception of the photon—a strictly 3D entity) exhibit both intrinsic (4D) spin and normal (3D) rotation. These motions, if aligned, allow for adherence of similarly charged entities leading to combinations and the complexity of tangible things.*

*A particle's size correlates with the intensity of force at its periphery—the smaller, the stronger. Neutrinos, defined by the weak interaction (roughly $1/1,000,000^{th}$ that of the strong), therefore, should be about $1,000,000$ times as large (as a proton); this helps to explain their ephemeral nature. Mass and frequency both define the same concept—proximity to a nuclear center—and increase (in tandem) the closer the objects they represent are to that core.*

*As most entities are but projections of a 4D reality, ever enlarging concentric 3D spheres (of energy—matter) define our existence. They help explain things from the tiniest transitory fragments to the immense vastness of space. Thus, to those spiritually inclined, this higher-dimensional reality, as it is the basis of all yet forever incomprehensible, often equates with God.*

# 36

# THE HYDROGEN ATOM

T HE BASIC BUILDING BLOCK OF NATURE, the hydrogen atom, is a 3D representation of a 4D sphere. Therefore, it presents as concentric, ever expanding globes of energy—matter.

The nucleus consists of dense, almost adherent spheres that serve as the event horizon for a 4D or black hole center. Surrounding this is a maelstrom or whirlpool of rotating spheres continually dragged toward that center. Most of the material is pulled in; some, however, barely escapes and is flung far out to coalesce as the delicate dust of our photonic seas.

Finally, a tangible essence persists, far enough away and rotating at a sufficient velocity to counter the centripetal pull of the black hole center (proton). This is our electron cloud, and due to its angular (circular) motion it has both electric and magnetic aspects.

Each 4D sphere represents a separate universe—all are entangled, all are one.

## Spheres Within Spheres

*Answer:* When you ask why things exist, you are really asking me to explain the basis of the hydrogen atom for, as you know, 99 percent of all matter is either hydrogen or helium. The easiest way to describe it is through our original concept of a three-dimensional surface on a fourth-dimensional sphere—our world. (Now, remember, due to the

— **247** —

curvature in every direction, there exists a continual increase in length—an ever-expanding redshift—and this uninterrupted stretching of space is what allows for visualization of the higher dimension.)

The hydrogen atom is the result of a similar fourth-dimensional globe (or hypersphere). This hypersphere exists far away in an unreachable plane; thus, its projection (our hydrogen atom) is quite small. We can only see a shadow, but, like our own world, it is understood by ever-increasing distances or redshifts. To simplify it somewhat, let me draw it as if we were but a two-dimensional surface world (Planeland) observing a remote three-dimensional globe. This may make it easier to visualize.

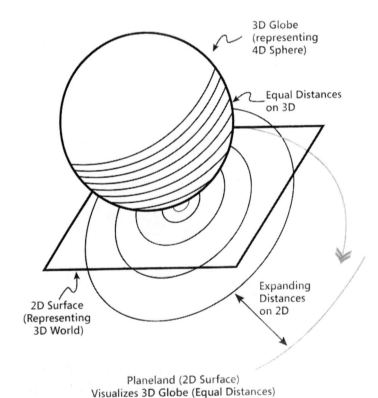

3D Globe (representing 4D Sphere)

Equal Distances on 3D

2D Surface (Representing 3D World)

Expanding Distances on 2D

Planeland (2D Surface)
Visualizes 3D Globe (Equal Distances)
as Expanding Distances

## The Proton Center

If we were simply two-dimensional inhabitants of Planeland, we would be wedded to this surface; we could not see or understand up or down. To us, these unknowable directions would be interpreted as just ever closer to a central point. So the more we attempted to comprehend this far-away, higher-dimensional object, the further concentrated would we sense that central core. We could never fully grasp this entity; we could only guess at its true character. Finally, all we could recognize would be a black hole—an object so dense that nothing, including light could escape.

**Question:** Let me see if I understand. The proton center of their hydrogen atom is sensed by Planelanders as nothing, as emptiness, as a black hole. But really they are attempting to visualize an incomprehensible higher-dimensional sphere (a three-dimensional globe). The closer they try, the more they gaze in that unfathomable direction, the more they are really looking inwardly and the denser it becomes.

**Answer:** You seem to grasp my ideas; therefore, to us (in our real three-dimensional world), the center is a black hole. Like our own universe it rotates at $c$ through the fourth dimension, through time. This rotation requires a centripetal force ($F = mv^2/r$) where $m$ is the mass of its surface, $v$ becomes $c$, and $r$ is an imaginary direction ($i$). Therefore, $F = mc^2/(i)$ or $(i)F = mc^2 = E$ *(energy)*.

Now, as we have already shown, each black hole or central proton represents an entire universe. The concentric, ever-expanding spheres of energy reach out to the very edges of existence. Each includes all—everything and everyone is entangled. Therefore, the energy of any one fourth-dimensional sphere contains that of all the others, and the total quantum energy of the universe turns out to be, more or less, $10^{120}$ times that of our three-dimensional world.

## The Electron Cloud

However, at a certain distance from our proton, a counterforce (of substance) is established; the distance and velocity of rotation allow a centrifugal expansion (equal to, and opposite of the centripetal pull), and material coalesces. Thus, an electron cloud forms, always at a uniform distance and rotating at a consistent velocity.

*Question:* So you are further describing our hydrogen atom. The center or nucleus is visualized by us as a black hole (a proton). If we closely examine its edge, we may see a very fine outline consisting of energy spheres closely compacted together; about this object is a maelstrom of circulating energy constantly pulled toward, and disappearing into, that black hole. However, finally, at some reasonable distance, a tangible mass forms with sufficient energy to counter this inward pull (an electron cloud). I guess visualizing it in two dimensions makes it easier to understand. How, then, do we translate it into our real or three-dimensional world?

*Answer:* Well, all we have to do is add height. Let me draw this.

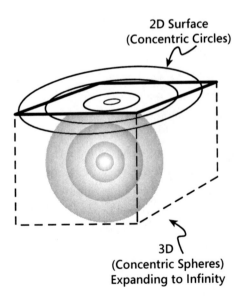

2D Surface
(Concentric Circles)

3D
(Concentric Spheres)
Expanding to Infinity

Therefore, to us, three-dimensional beings, this fourth-dimensional sphere is an ever-expanding set of concentric energy shells to the very limits of reality. However, they appear tangible (to have substance) at both the nuclear edge and the electron cloud.

*Question:* So electron clouds become the palpable exteriors, the concrete manifestations of how fourth-dimensional spheres are understood or visualized in three dimensions. They need always exist as rotating clouds about nuclear centers (as hydrogen atoms) and they must all be the same size (or distance from that core) as that is simply where they can form.

## The Photonic Sea

*Answer:* Yes. Now, remember, we are visualizing the fourth-dimensional sphere in our terms. We note a center, we see an electron cloud; between is an unstable vortex of energy shells constantly pulled toward that core. However, some small percentage barely escapes (flies by that center) and is expelled (to orbit far beyond the electron cloud). That material coalesces into very fine particles, the delicate dust of our photonic sea. These particles (these photons) have mass; in fact, they give rise to our so-called dark matter.

*Question:* So you envisage three essential parts to every hydrogen atom: the proton center, the electron cloud, and the photonic sea. All atoms, since all are but combinations of hydrogen, therefore, should be similar.

*Answer:* Yes, all atoms contain similar entities—a nucleus, an electron cloud, and photonic outer sea. They are merely manifestations of the simple hydrogen atom, initially fused under the immense heat and pressure of stars to be finally formed in supernovas.

## Rotation

*Question:* But what about all the characteristics of the electron? It orbits about the nucleus, it has magnetic properties (we always discuss electromagnetism as one concept), and it has some kind of intrinsic spin. How do you explain these attributes?

*Answer:* At the distance from the black hole center (or proton) where an electron is found, there are multiple concentric energy spheres that, to keep from being pulled inwardly, rotate countering that force. This is its angular or circular motion. But remember, each concentric sphere rotates as a single entity; thus, each moves fastest at its equator and slowest at its poles. This causes a change in shape—a midline bulge and a flattened top and bottom.

As we travel from equator to pole, the sphere becomes less and less able to counter the pull of the black hole nucleus; thus, it is drawn to the core, just like that inner maelstrom of energy, and, if barely missing, is flung out to orbit in a north/south or vertical orientation.

Therefore, we have both horizontal and vertical motions; the horizontal described as electric, and the vertical as magnetic. As already noted, a north pole, due to the direction of its motions, conforms to a south pole of another entity (north attracts south) but opposes, or interferes with, a like pole (north repels north). Furthermore, as already stated, a magnetic monopole cannot exist, as a black hole cannot be sectioned; if it were it would instantly reform or pull the entire universe into itself.

*Question:* But what about other motions? Isn't there some intrinsic circular movement that cannot be explained, that we call spin?

*Answer:* To understand these, we must once again visualize the fourth dimension. Let us now attempt to describe these movements through our plane of existence, from virtual to real and then we will try to decipher spin.

# 37

# THE DIVIDE—REAL/VIRTUAL

I *F ONE WERE* 2D *(NO HEIGHT), a* 3D *globe would present as ever-expanding circles starting from a central spot and continuing to infinity. In a similar fashion, given our 3D world, a 4D orb (hypersphere) appears as ever-enlarging, limitless, concentric spheres of energy—matter.*

*Individual particles are only different aspects of these hyperspheres; the hydrogen atom, however, represents one in its entirety. Protons are attached to the interior of the 3D divide (a divide that can only be understood from a 4D perspective) whereas neutrons are attached to its exterior. When a neutron is freed from its deep, secure spot in a nucleus, it emerges to our side, and, since it represents an entire hypersphere (just as does a proton), it brings along its concentric covers (electron and neutrino).*

*As these covering shells emerge to our plane, we first see their inside (antiparticle), then their outside (regular particle). Photons, as they are merely the traces of previously existing shells, are found only in our tangible 3D (never traveling through the divide); thus, photons have no corresponding antiparticles.*

## Hypersphere

Let us try to do the impossible: visualize a fourth-dimensional sphere (hypersphere). Remember, if we were only two-dimensional and lived in Planeland (our flat, fantasy realm), a three-dimensional sphere would appear as concentric, ever-enlarging circles radiating to infinity. A knowledgeable two-dimensional inhabitant could attempt to construct a globe from these rings but would have great difficulty placing them one atop another, since that direction would not exist. However, what the Planelander could envisage was that both the outsides and the insides (or anti-rings) were equally present, and that smaller ones fit within larger ones all the way to a black hole center. If we were to draw it, we would get the following:

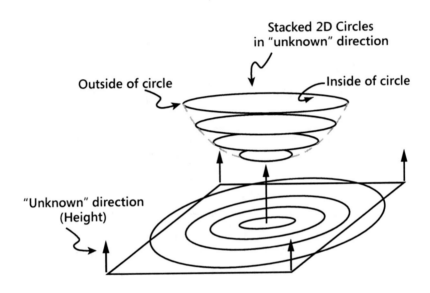

Stacked 2D Circles
in "unknown" direction

Outside of circle

Inside of circle

"Unknown" direction
(Height)

Therefore, we get stacked circles—to us obviously a three-dimensional sphere, but to the Planelander simply ever-widening discs.

In a like manner, a hypersphere presents to us as ever-enlarging concentric spheres to the very edges of our universe. We cannot draw it, since we have no fourth-dimensional perspective; we are unable to imagine the inside and outside of our world. But, just as a two-dimensional denizen can *guess* at a third dimension, we can similarly make an attempt at visualization. Thus, a hypersphere can be seen as separate, ever-enlarging globes, with anti-globes, in an unknowable direction.

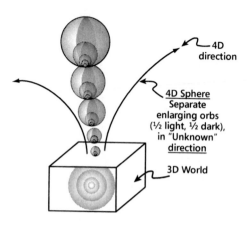

These entities are all present at the same time—they are all one complex object. Thus, another projection would be of transparent orbs, with anti-orbs, one behind the other, all potentially visible to the viewer. They would all appear as separate, yet part of the same body. They would all touch the surface of our world at the same spot, their centers aligned as a straight path into that higher dimension. Let me attempt to picture it.

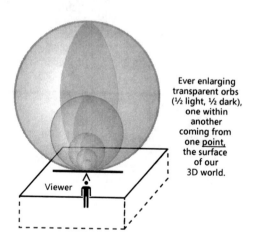

*Question:* Okay, I am beginning to see what you are trying to show. Each part of the surface of the hypersphere to us is a three-dimensional ball. As we attempt to see the entirety, we see ever-expanding globes, with anti-globes, nestled within one another, attached to a single point. In our world, that becomes the very center of a black hole. That spot, though, is really the attachment of this higher dimension to our plane of existence.

## Neutron

*Answer:* You seem to grasp what I am attempting to describe. Let me try to put this into more concrete terms. Remember the discussion of a neutron, an entity that only exists within the nucleus of some atom. Once freed, it changes into a proton, an electron, and an antineutrino. If I were to draw it using a fourth-dimensional perspective, perhaps our ideas might make sense.

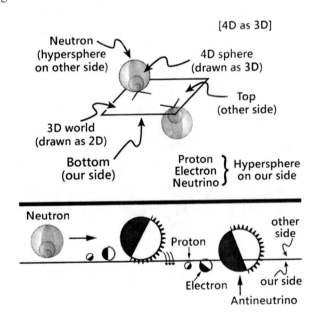

We only see ½ of entity. If sphere is
mainly on our side "particle", if mainly
on other side "antiparticle"

We are drawing a neutron as a potential particle—it can only exist in the center of a nucleus; once in the actual world, it changes into a proton, an electron, and an antineutrino. Thus, we are stating that it is a hypersphere, but on the *other* side of our three-dimensional plane; it only potentially exists in our world. However, it is attached to the surface of our world (on the opposite aspect); it is within the nucleus of an atom.

Upon entering our realm, the real world, and leaving the nucleus, it must take on its true characteristics; it must be seen as concentric spheres ever enlarging to infinity. Thus we find a proton or concentrated center and its counterbalance, an electron cloud. These are the tangible parts, the elements that give substance to our entity.

## Neutrino

We also find a neutrino—actually an antineutrino. This particle has minimal mass and is difficult to describe. It interacts with almost nothing and can easily travel through many light-years of matter. It simply is the surface of our hypersphere farther (fourth-dimensionally) from the point of attachment, thus significantly increased in size but considerably decreased in density.

The reason it presents as an antiparticle is that, in the time it takes for a hypersphere, sensed as a neutron, to pass through the surface of our world and become a proton (about 15 minutes), only that aspect appearing as an electron can also clear. The neutrino, being much larger, can only partially enter our universe; it is less than halfway in; thus, it presents inside-out (as an antiparticle).

## Virtual, Potential, And Real

*Question:* So you are stating that a neutron is a potential entity consisting of a proton, an electron, and a neutrino, in essence a hydrogen atom attached to the other side of the plane of existence. When it enters our world, when it disengages from the center of the nucleus and crosses this plane, we envisage real particles—a proton, an electron, and an antineutrino.

But I thought that potential or virtual particles just pop into existence with their antiparticles and disappear almost instantaneously with a great deal of energy. Why then would the neutron be such an entity? It certainly always exists, even if only within a nucleus, and when it exits this home it lasts for palpable minutes—certainly more than just a fraction of a microsecond.

*Answer:* The reason is that the nucleus of an atom is really, at its innermost center or fourth-dimensional aspect, the home of potential objects—those just outside our existence. However, as the nucleus itself is tangible, all parts of it (real and potential) make up its heft (its mass). Thus, the protons and neutrons within a nucleus account for its mass, but the protons are on our side (the inside) and the neutrons are on the other side (the outside) of the realm. They are both attached to the surface of our three-dimensional existence, simply on opposite borders (fourth-dimensionally).

*Question:* I see, the inside is what is real, the outside virtual or potential, but if both are within a nucleus, both are measurable—both are attached (albeit on opposite surfaces) and have mass. I guess the other many, many virtual particles (I think you claimed they were 100,000 trillion, trillion, trillion times more plentiful than real ones) are not connected to our plane of existence, thus have no heft—they leave no footprint. When found, they are but briefly entering to quickly leave with a bang.

*Answer:* You are correct. Practically all particles are virtual and appear to come and go almost instantaneously. They have no determinable mass except for the moment they appear. However, their energy is present in the great vacuum of our universe. They are why the theoretical quantum energy is, more or less, $10^{120}$ times as great as that actually found.

## Proton To Neutron—Inside To Outside

*Question:* Since a neutron cannot exist by itself in our real world but becomes a proton, electron, and antineutrino, what happens when a proton changes over to a neutron within some nucleus?

*Answer:* The opposite occurs. A hypersphere attached to our surface (inside) moves through the plane of existence to the other surface (outside). But, just as before, since it takes time, some parts are not fully through when the proton initially changes (becomes a potential entity, a neutron). Thus, a proton can become a neutron, but it is accompanied by an antielectron (positron) and a neutrino. Again, let me draw this.

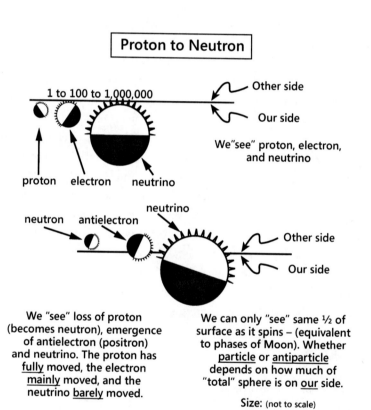

## Proton to Neutron

1 to 100 to 1,000,000

Other side

Our side

We"see" proton, electron, and neutrino

proton    electron    neutrino

neutrino

neutron    antielectron

Other side

Our side

We "see" loss of proton (becomes neutron), emergence of antielectron (positron) and neutrino. The proton has fully moved, the electron mainly moved, and the neutrino barely moved.

We can only "see" same ½ of surface as it spins – (equivalent to phases of Moon). Whether particle or antiparticle depends on how much of "total" sphere is on our side.

Size: (not to scale)

The hypersphere in the real world—our equivalent of a hydrogen atom (proton, electron, and neutrino)—moves through the plane, and we find, as it has not fully entered the virtual realm, a neutron, antielectron, and a neutrino. The electron is still somewhat (less than one-half) in the real world, and the neutrino is yet mainly on our side of the plane.

## The Missing Anti-Photon

*Question:* But are you not being a little loose in your definitions of a hydrogen atom? Previously you said it had a proton, an electron, and a surrounding photonic sea. Where is this sea? Have you not mixed it up with your concept of a neutrino?

*Answer:* No, the photonic sea is a strictly third-dimensional construct formed by the breakage and subsequent narrow escape of remnants of unstable energy spheres (those pulled toward the black-hole center by their close proximity and lack of sufficient counterforce). The remaining fine particles (photons), dispersed through three-dimensional space, are not really intrinsic parts of a hypersphere or fourth-dimensional globe. The neutrino, on the other hand, is composed simply of energy orbs extending much farther from the initial point of attachment (to our world) and moves in tandem with this great fourth-dimensional entity.

In fact, the photon, as it is strictly a third-dimensional object, cannot travel through the plane of existence to the virtual side. Thus, it cannot become an antiparticle (anti-photon), nor, as we will shortly see, is it affected by the "exclusion principle," and, a full rotation takes merely 360°. Modern physics has attempted to solve this problem (lack of a distinct anti-entity) by claiming that the photon is its own antiparticle. In reality, there is none, there simply are photons and they stay on our side of the fourth-dimensional divide.

## Two Sides Of The Divide

*Question:* So a proton–neutron exchange actually represents the movement of a hypersphere onto the virtual side of the plane, and, when the nucleon (proton) has fully exited, the electron still has some time to move, and the neutrino, as it is so much larger, has merely begun.

*Answer:* You are really getting the gist of my ideas. There is a real world, our side or the inside of the plane of three-dimensional existence. There is a virtual world, either attached to this divide, the potential realm (within a nucleus), or entirely separate (unmoored and freely dispersed in the great vacuum of the universe). Then there is a halfway as a particle traverses this partition. If over one-half is outside, it is an antiparticle; if over one-half is inside, it is a real particle. It is simply the same entity, seen on the outside or inside—fourth-dimensionally.

Lastly, since we are describing neutrons as potential hydrogen atoms, stuck to the "other" surface but registering on "our" side (as massive, intra-nuclear material), the H/He ratio (75/25), previously mentioned and thought to be a result of the Big Bang (the residue of energetic photons and nucleons—protons and neutrons—in a $10^9/1$ proportion), may, instead, be seen simply as what neutrons are most capable of achieving. They can only exist if within a nucleus, and because helium (two protons with two neutrons) is the most stable simple element (after hydrogen), it is, by far, their most likely home.

Thus, since hyperspheres, attached to the outer surface (within nuclei), exist, as neutrons, helium becomes a very frequent arrangement. The cauldron of a star's interior is not needed for most of helium's manufacture but does become necessary (with the explosive power of a supernova) for the great percentage of more complex elements.

**Question:** So you are finally attempting a logical basis for the H/He ratio, long considered a fundamental tenet of the Big Bang. If I were to accept your concepts, then there really is no need for current theory.

**Answer:** Yes, if you agree with my thesis—of multiple, spinning hyperspheres—then, I think, a simpler picture emerges. It allows for a clearer understanding of redshifts, CMBR, and the H/He ratio, for a firmer basis for today's beliefs. Now, as we both vaguely understand the hypersphere let us try to discuss other aspects of its existence, specifically its motion and spin through the higher dimension.

# 38

## CIRCULAR MOTION

MOST PARTICLES HAVE TWO CIRCULAR MOTIONS, both 4D (intrinsic spin) and 3D (rotation). Intrinsic spin (as it is 4D) is only noted when the object crosses our plane of existence (enters our 3D). Although like particles are thought to repel, the Pauli exclusion principle allows those with opposing spin to adhere, or coexist. In these cases, both bodies enter our plane in the same direction (thus, no interference). Zitterbewegung (tremulous motion) is based on intrinsic spin; it is seen as the entity rapidly and continuously exits, then re-enters our world.

Bosons, carriers of force (e.g., photons) have full-integer spin, whereas fermions, material particles (such as electrons) have half-integer spin. This generally means that bosons spin around once ($360°$) before returning to their original configurations, whereas fermions must spin twice ($720°$) to re-establish themselves. Quantum physics explains this mathematically, but, to fully comprehend it, one must visualize the 4D.

Rotation, 3D motion, also allows for like objects to adhere if their movements are opposed (clockwise vs. counterclockwise) because, when approaching one another, just as with intrinsic spin, their surfaces would be traveling in the same direction. 3D rotation, in its vertical and horizontal aspects, helps to explain the octet rule (the 8-electron configuration of greatest stability).

Finally, centripetal force ($mv^2 / r$—the basis of the four fundamental forces) proportionally decreases with distance, but electromagnetism and gravity decline faster (with distance squared), and the strong nuclear force dissipates precipitously (after leaving the confines of a nucleon) Although all represent the same concept, they appear different due to placement in the third and fourth dimensions.

## Intrinsic Spin

The hypersphere is the basis of our universe. The entire knowable world covers just less than one-quarter of the surface of some great globe. All the many and varied things that we see, feel, or understand to exist are composed of these fourth-dimensional objects. As we are cognizant of only three dimensions, we can only really know the exterior of these entities—their *outline* in our world. However, we can infer other aspects if we delve into the higher dimension.

Remember, the essence of time is the movement, at the speed of light, of our great orb toward the fourth dimension. It gives us a recognizable past, a fleeting present, and an ever-unfolding future. All other such entities also move toward this higher dimension at $c$. This motion is what makes for the intrinsic spin of all particles.

## Pauli Exclusion Principle

Wolfgang Pauli, a brilliant theoretician who first conceived of the neutrino, described a fundamental concept (later named, in his honor, the *Pauli exclusion principle*), which allows that, among particles comprising matter, only those with unequal quantum characteristics can occupy the same physical space. Various numbers are given to identify different quantum aspects of these specific particles, and, as long as these integers are not comparable, similar entities can coexist. However, if they are equivalent, peaceful cohabitation is impossible; one of them is the intrinsic spin of that object.

In an electron, this spin is noted as plus or minus one-half. Thus, a pair of plus one-half electrons cannot be in the same space (if their other values are similar), whereas a plus and minus couple can. In quantum physics, spin is not a concrete or real motion; it is just a concept that either does or does not allow for harmony.

*Question:* Well, if quantum perceptions are fourth-dimensional, why not dissect spin? Perhaps you can then give a mechanical, albeit other-worldly, explanation.

*Answer:* It is actually not that mysterious once you move to a higher plane. Objects can coexist if, upon contact, they are moving in the same direction; they cannot if their movements are opposed. In the case of two electrons, each would be part of a distinct hypersphere spinning toward that higher dimension at *c*. When they touch, if their spins allow the same direction (fourth-dimensionally), these objects can interact or co-join; however, if their spins lead to opposing directions, they interfere—they cannot come into close proximity. Let me draw this.

## Spin

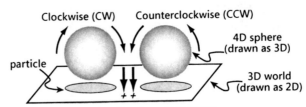

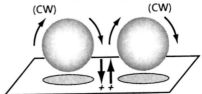

Spin in opposite directions (CW/CCW)
upon contact, movement is in <u>same</u> direction
therefore, objects adhere or co-exist.

Spin in same directions both CW or CCW
upon contact, movement is in <u>opposite</u> directions
therefore, objects interfere; they <u>cannot</u> co-exist.

What we have drawn is our three-dimensional plane as a two-dimensional surface and higher-dimensional spheres as three-dimensional globes (omitting light and dark halves for clarity). When they spin contrary to one another (clockwise *vs.* counterclockwise), they come together traveling in the same direction and can adhere. When they spin in the same way (both either clockwise or counterclockwise), they approach from opposite directions and interfere—they cannot coexist.

**Question:** I see what you are trying to explain. The Pauli exclusion principle describes motion through a higher dimension leading to changes, adherence or interference, in our world. Thus, it really is a mechanical construct not just a mathematical notion.

## Zitterbewegung

**Answer:** You do grasp my ideas. Wolfgang Pauli obviously understood the math, he just did not visualize the actual effect. Another interesting concept is what in German is called *zitterbewegung*, or tremulous motion. All particles have a fine cyclical jitter, again discovered mathematically. It occurs at a very high frequency and is real but not well understood.

**Question:** I love that word—*zitterbewegung*—although I am not sure how it is pronounced. How would you explain it and attempt its visualization?

**Answer:** Again, using our fourth-dimensional concept of spin at *c*, we can show what is occurring. There is essentially a very rapid loss, then re-emergence, of what one considers a three-dimensional particle as it spins in the fourth dimension. Let me again draw our concept.

### Zitterbewegung (ZBW)

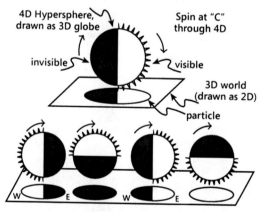

| ZBW is the jitter from West to East (or vice versa) as the particle disappears and reappears while the hypersphere spins at "C" through the 4D. | As we always see but one side, it is analogous to the phases of the Moon |

You can see from the drawing that, as the particle spins clockwise or counterclockwise (into the fourth dimension), it continuously leaves, then re-enters, our plane of existence. This constant disappearance and reappearance is the jitter or tremulousness of *zitterbewegung* and, as the object spins at *c*, it occurs at an extraordinarily high frequency (in the case of an electron, at 750 million trillion cycles per second).

## Fermions, Bosons, and Mirror Images

**Question:** So the Pauli exclusion principle and *zitterbewegung* can be visualized if one uses the fourth dimension. Before, you mentioned the electron's spin was plus or minus one-half. Why not elaborate on that now?

**Answer:** Let me make an attempt. Particles are thought to be roughly divisible into two major groups. There are those that constitute matter (fermions—named after Enrico Fermi, an important Italian physicist), and those that are believed to carry force (bosons—named for Satyendra N. Bose, an influential Indian scientist).

When discussing the latter (bosons), they are described as having integer spin (0, 1, 2, . . .), whereas the former (fermions) are noted to have half-integer spin (1/2, 3/2, . . .). What this generally means is that, if a boson (e.g., a photon) is rotated completely around (one full $360°$ turn), it looks the same as before. A fermion (an electron, in this case), however, needs to be rotated two times ($720°$) before it re-establishes itself.

**Question:** Well, how do you explain this? Why would something have to turn around twice to look the same? Doesn't every object have a front and back, and, when fully spun ($360°$), return to its original configuration?

**Answer:** Just as with our prior examples, to fully visualize this one must use a fourth-dimensional perspective. Remember, most particles are lower-dimensional reflections of constantly spinning higher-dimensional orbs. We always observe merely the same one-half of that sphere (the "anti-half" is not visible), and, as that segment's image enters and leaves our plane, we see phases (similar to those of the moon)——our so-called wave–particle duality.

But, although the particle gets smaller and disappears from our surface universe, it has not actually vanished from its higher plane. It has just spun around and will reappear, to us, again, as an enlarging sphere. However, its momentary absence really can be understood as a complete reversal in our world. It has traveled through itself, existing now as a potential antiparticle (its inverse). Let me attempt to draw this.

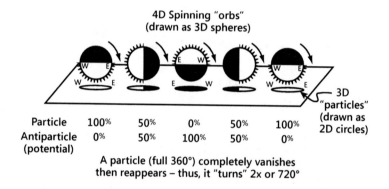

4D Spinning "orbs"
(drawn as 3D spheres)

3D "particles" (drawn as 2D circles)

| Particle | 100% | 50% | 0% | 50% | 100% |
|---|---|---|---|---|---|
| Antiparticle (potential) | 0% | 50% | 100% | 50% | 0% |

A particle (full 360°) completely vanishes
then reappears – thus, it "turns" 2x or 720°

As the particle appears to vanish (to the west), to finally, completely, disappear, it has really flipped (east has become west, and vice-versa). However, we can no longer visualize this entity. It has rotated fully, but through itself and our plane of existence. It has become its inverse, its antiparticle, but, as it is on the "other" side of a three-dimensional surface, it cannot be seen; it is only "potentially" present.

The object, then, begins to reappear (in our drawing always from the east) and finally returns to its original aspect. Thus, in reality, it has completely passed through itself twice: the first time to its inverse (invisible to us, as only potentially present in our lower realm), and then back to its initial configuration. It, therefore, has rotated fully two times (inside-out, then back again, outside-in), or 720$^0$.

**Question:** Your concept is not really that easy to follow, but I think that I understand it. Perhaps I can simplify it somewhat with a more concrete example. If your quantum particle were, in actuality, a macro object, a person, for instance, then it may become clearer. Let me try using mirror images.

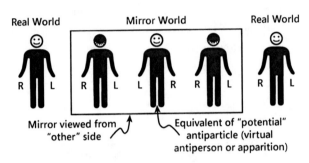

Real World       Mirror World       Real World

Mirror viewed from "other" side

Equivalent of "potential" antiparticle (virtual antiperson or apparition)

The "person" turns 2 full rotations or 720°

In the drawing, you can see that the mirror image reverses sides (with the individual) right becomes left and left, right. If that image were somehow, then, to turn, by itself, facing the back or "other" side of our mirror (obviously not possible in the real world), it would be the same as your potential antiparticle—a virtual anti-person, or apparition. If it once again pivoted and then exited the mirror the individual would resume his or her original aspect. In so doing, the person would have fully rotated two times, or 720°.

*Answer:* Your drawing does appear to be somewhat more intuitive; however, for your "person" to match my example, he or she would really need to be, at all times, two entities both real and potential; the latter, or doppelganger, only coming to "life" in a virtual "mirror-world." Remember, you are still using a two-dimensional depiction of a full-bodied individual, and only the "magical" mirror allows you to see the front and back of this supposed person. If, as in your drawing, only a front exists, when turning, he or she is really stepping into a higher (or third) plane, and, thus, like a potential antiparticle, completely vanishing.

What I have been trying to show is that a fermion, or material entity, as it is a reflection of a fourth-dimensional orb, is always composed of a particle and its latent antiparticle (the outside and inside, so to speak, of a higher-dimensional object). Therefore, its rotation is actually that of two separate $360^0$ bodies (one real and palpable, one dormant and intangible) and must encompass $720°$.

What is generally believed to be a boson, on the other hand, a carrier of force (in this case, a photon), is actually a strictly third-dimensional item. Thus, it has no "other" or fourth-dimensional side—no anti-boson (or anti-photon)—and, being but a solitary particle, rotates fully in $360°$. Finally, the Pauli exclusion principle only relates to fermions (entities that are reflections of fourth-dimensional spinning spheres), as it is only in the higher realm that this motion occurs. Bosons (strictly third-dimensional elements) have no "other-worldly" spin, they simply rotate in our plane, and, accordingly, no exclusion pertains.

Let us now return to the "actual" world—lower-dimensional reality. In this world, our electron cloud has an entirely different direction of motion, a three-dimensional rotation completely separate from fourth-dimensional spin. This rotation, when opposed (just as with the Pauli exclusion principle), is a major reason for the adherence of like-charged particles (electrons). It also leads to magnetism (as already shown) and allows (with opposing fourth-dimensional spin) for atoms to form together into molecules—leading to all we behold.

## Octet Rule

*Question:* I vaguely remember something from my high school chemistry called the *octet rule*; that the greatest stability occurs when there are eight valence electrons. Do the motions discussed have anything to do with this?

*Answer:* I am glad that you bring it up. Let me try to explain its basis. The octet rule was formulated by two very influential chemists of the early 1900s—Gilbert N. Lewis and Irving Langmuir. Essentially, the most solidity occurs with eight electrons in the outer shell of an element.

As we have been stating, each electron has two movements—a third-dimensional rotation, and a fourth-dimensional spin. When these line up, we get permanence; spin allows for coexistence, rotation for adherence.

Thus, the octet, or eight-cornered arrangement, is the steadiest as all motions align. Let me attempt to draw this.

## OCTET RULE

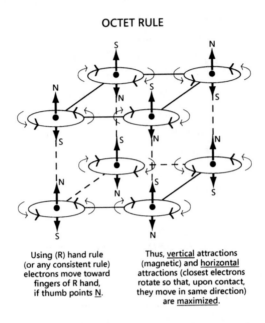

Using (R) hand rule
(or any consistent rule)
electrons move toward
fingers of R hand,
if thumb points <u>N</u>.

Thus, <u>vertical</u> attractions
(magnetic) and <u>horizontal</u>
attractions (closest electrons
rotate so that, upon contact,
they move in same direction)
are <u>maximized</u>.

The closest partners of the cube move in tandem with one another; thus, they adhere, and the reasons are found in both the higher and lower dimensions. Hence, the octet rule (eight electrons at the corners of a cube) accounts for maximum strength.

## Charge, Distance, and Strength

*Question:* I see; you are using their movements (spin and rotation) to allow like particles to coexist and adhere. The negativity of electrons, supposedly causing mutual repulsion, then, is a motion phenomenon and can be overcome by appropriate placement.

*Answer:* That is correct. We have always said that the concept of positive and negative is simply the force—counterforce caused by the pull toward the center. The real reason for repulsion is cyclical movement—rotation in three dimensions, and spin in four.

**Question:** So if attraction and repulsion are merely circular motions, what then does the charge of a particle really signify? Obviously it must describe something.

**Answer:** Charge is the measure of the *direction* of the force on an object. When sensed from the outside, it is deemed positive, or pulling toward a center. When detected from the inside, it is considered negative, or pushing away from a core. Do not forget, an antiparticle is the same as its particle, only visualized (in fourth-dimensional terms) from the inside out. Therefore, charge simply tells us from which side we are observing that entity.

**Question:** So if charge tells us from whence a force originates, why does that force sometimes decrease solely with distance (centripetal attraction), at other times much faster, with distance squared (electromagnetism and gravity), and, finally, as with the strong force, precipitously, over a very short expanse?

**Answer:** Let me try to explain. When mentioning centripetal pull, we are noting the basis of the universal force, a fourth-dimensional attraction seen by rotation toward time of our world on a great higher-dimensional globe. This tug is the cause of all force; it encompasses everything. It gives form to different particles. The closer these entities are to the source (the fourth-dimensional center), the denser they appear and the greater their pull; but, since it is centripetal ($mv^2/r$), it is simply based on the direct distance, or radius, from that core.

Gravity and electromagnetism, on the other hand, maintain, and are felt within, the three dimensions of our world. Each radiates out from a central source or particle in all possible directions—essentially as a sphere. The force noted at any set distance, therefore, is spread, or dissipated upon, the surface of this sphere ($4\pi r^2$, its formula); thus, it is proportional to that distance, or radius squared. Finally, the strong nuclear force is really the equivalent of gravity within an entirely separate universe; consequently, it only functions inside the confines of that domain. Therefore, its power rapidly disappears after $10^{-15}$m (the approximate radius of a nucleon).

Let us expand on these concepts in the next chapter; perhaps, by using specific examples, they may become clearer.

# 39

# EVANESCENT PARTICLES

T*HE MUON AND TAU ARE TWO of many transitory particles momentarily presenting after violent collisions. They are similar to, but lack the stability of, recognizable entities (protons and electrons). Their mass is proportional to their proximity to the nucleus (black-hole core), and their charge merely connotes the counterforce of inertia.*

*A particle's size correlates with the intensity of force noted at its surface: The smaller in size, the greater the force. Thus, the strong force at the periphery of a proton (approximate diameter $10^{-15}$ m) is about 100 times that of electromagnetism, the force at the edge of an electron (Compton wavelength, or size of an electron, around $10^{-13}$ m).*

*The neutrino, defined by the weak interaction (roughly $1/1,000,000^{th}$ the strength of the strong force), thus, should be 1,000,000 times as large (about $10^{-9}$ m). This allows it to ephemerally pass through literally light-years of denser matter. Our universe (held together by gravity), therefore, if viewed as an immense particle, is $10^{41}$ times as large as a proton but with a strength of adherence $1/10^{41}$ (or $10^{-41}$) as strong.*

## Muon

When scientists send particles energetically colliding into one another, they discover many new, unusual, and very short lived entities. One of the more stable such objects is the muon. It has a negative charge, just like the electron, only it is over 200 times as massive. One of particle physics' important pioneers, Isidor I. Rabi, when first noting it, face-tiously quipped, "Who ordered *that?*" as it was so strange and out of place in the theoretical framework of that time.

**Question:** Well, why then do you bring it up? What purpose does it serve, other than to confuse us?

**Answer:** The reason I discuss it is that the muon, like so many other potential or virtual objects (particles that last but for the shortest imaginable time and are found only with the disintegration of stable entities), fits the overall concept of force and counterforce. The muon is simply material between a proton and electron (that active mael-strom of energy spheres constantly pulled toward a black-hole center) that has been disrupted by a violent impact (often a particle physicist's experiment) and, for a fleeting instant, appears as a distinct, tangible entity.

It has a negative charge equal to that of an electron, yet its mass, as it is nearer the center, is quite a bit larger. Remember, the energy spheres of which it is composed are closer together, making it more compact or denser (greater in mass) than a more distant electron.

**Question:** So this, and many other weird particles found in colli-sions, only come into existence because of these disasters. Without dis-ruptions, they would remain potential or virtual. It is only the high-energy experiments or cosmic ray bombardments that rouse them and allow their quick entrance and exit from our world.

**Answer:** That is correct; they would lie hidden, as they are not stable in our three-dimensional plane. Once brought forth, they take on the characteristics of more established players with similar charge depending on whether they are sensed as part of the central pulling force or coun-tering it (i.e., visualized from the outside or inside). Their mass is simply related to their distance from that black-hole center. But the charge of these entities (plus or minus) as seen in protons and electrons is due to force and counterforce.

## Tau

Let me bring up another entity that has a fleeting existence, even shorter than a muon's, but, like it, has the same negative charge—the tau particle. The tau, similarly to the muon, only comes into existence after some catastrophic event—breakage of a stable atom by an energetic cosmic ray or a head-on collision in a particle accelerator. Although it has the same charge as an electron, it is much more massive (over 3,500 times or about twice the heft of a proton).

As it is so concentrated, it would reside even closer to that black-hole center than our proton; but, unlike the proton, it carries a negative charge. Therefore, it exists, or is understood, as a counterforce similar to an electron (it is visualized from the inside). However, it is not stable; it decays rapidly, and, like the freed neutron or vanishing muon, one of the byproducts is a neutrino.

So just as a neutron leads to a hydrogen atom with proton, electron, and neutrino (or antineutrino, if you prefer), the tau also decays to products including a neutrino. Because of the establishment of the neutrino, these decays are felt to be due to the weak nuclear force (often simply called the weak force or interaction).

## Strength And Distance (From The Center)

*Question:* We have discussed the universality of force several times. Since you are now bringing up the weak force as a specific entity, why not elaborate further on your ideas?

*Answer:* Yes, let me try to explain. All the forces are really one: different aspects of a centripetal pull set up by the fourth-dimensional rotation of mass at $c$ (the speed of light). This force is, to us, tugging in an imaginary direction ($i$) toward the center of that unknowable, higher-dimensional globe. But the force is usually understood and written as:

$$F = mv^2/r.$$

Now, $m$ is the mass of an object in three-dimensional terms (density of energy spheres, the greater the closer to the central point), and $r$ is the radius of a particle as it appears in our world. However, $v$ is really $c$, its velocity in the fourth dimension, as it is rotating and establishing this force in the higher realm. Therefore, if:

$$F = mv^2/r, \text{ then } F = mc^2/r, \text{ or}$$

$F$ is proportional to the inverse of $r$ ($1/r$); thus, the larger the radius ($r$), the smaller the force ($F$).

**Question:** But previously you were claiming that the radius ($r$) was imaginary ($i$). Aren't you playing a bit fast and loose with your definitions? How can you now use a real value instead of an imaginary concept if we are discussing an unknowable dimension?

**Answer:** I am not trying to confuse or confound you; I am merely using the size of the particle as it appears in our world. Remember, the fourth direction exists, only it cannot be comprehended by our limited senses. All objects have actual size in that dimension. We visualize their shadows, but these proportions (these ghosts of a higher reality) equate to appearances in that other plane.

Now the strong force is seen at the edge of the proton whose radius is approximately $10^{-15}$ meters. The electromagnetic force is noted at the edge of the electron; its radius (Compton wavelength or smallest possible size) is roughly $10^{-13}$ meters (100 times as big as a proton). Thus, the electromagnetic force is about $1/100^{th}$ the strength found at the proton's edge.

## Bohr Radius Vs. Compton Wavelength

**Question:** Just before, you were discussing the Bohr radius of a hydrogen atom; now you are noting the Compton wavelength of an electron. What is the difference between these two concepts?

**Answer:** The Bohr radius is the most likely orbit of the electron circling a hydrogen atom (about 5 x $10^{-11}$ m); it describes a three-dimensional configuration. The Compton wavelength is the smallest size of any particle, in this case an electron (approximately 4 x $10^{-13}$ m), and it is how a fourth-dimensional entity is understood in our world. It defines and establishes electromagnetic force. The difference in size is a factor of around 100 (137 to be more exact—the *fine structure constant*).

**Question:** Okay, I see the difference; one is a three-dimensional configuration (Bohr), the other is really part of a fourth-dimensional hypersphere (Compton).

## Neutrino And The Weak Force

**Answer:** Yes, that is a reasonable way of understanding them. To get back to the weak interaction, it has about one-millionth the intensity of the strong force; thus, the radius of the particle that it creates (in our world) would be around one million times as great as that of the proton (or $10^{-9}$ meters). Since the weak force is closely aligned with the establishment of neutrinos, the size of a neutrino should be, if this analogy holds, one million times that of a proton, or ten thousand times that of an electron.

At this great distance, its energy spheres, due to the redshift, are significantly widened. It is much less compact; thus, its heft or mass is markedly diminished. Since it is so less massive, for all intents and purposes it is ephemeral and can easily pass through literally light-years of other matter.

Its wavelength is so much greater than that of a proton (or even an electron) that most interactions with other entities, therefore, are exceedingly rare. This ability to traverse what we consider solid is really not that unusual; it parallels an infrared wavelength's transparency to interstellar dust or a light wave's penetrance of glass. (In both cases the greater wavelengths of those electromagnetic disturbances allows for their easy travel through otherwise thicker, or denser, material.)

*Question:* So the weak force defines the neutrino (a wisp of a particle), which can easily move through what to us appears solid, as its wavelength (or width of energy spheres in three dimensions) is much greater than the (dense) material through which it passes?

*Answer:* Yes, all forces are similar; all are aspects of that fourth-dimensional rotation at $c$ seen in our realm as the pull toward a center. The larger that sphere, the more diffuse (less dense) it appears, and the feebler is the force. The weak interaction is one million times less potent than the strong; thus, the entity defined is one million times as large.

*Question:* I guess you can throw gravity into this mix. The universe is approximately $10^{26}$ m in radius; thus, it is $10^{41}$ times as great as a proton, and the force holding it together (gravity) is $10^{-41}$ times as potent.

*Answer:* I fully agree; the fundamental forces are merely different aspects of that centripetal pull. Gravity, therefore, is only the remnant of this attraction diluted by the great expanse of our world.

## Neutrino Flavors

*Question:* We were speaking of neutrinos, those barely existent entities, and I remember reading somewhere that there are different types, or "flavors". Do they all have similar characteristics?

*Answer:* I am always surprised by your fount of knowledge. Yes, there are three distinct kinds (or flavors) of neutrinos—one each for the electron, muon, and tau. They appear to be interchangeable to a limited extent. However, their natures are really not that different, and they are all affected by the weak force. In a sense they define the outer reaches of our negatively charged (or counter) particles.

They are considered to be neutral, not influenced by electromagnetism. However, this is not really understood, and I would certainly not at all be surprised if they had a very weak charge (counterforce) of their own. But we only find them in unsettled situations (either in neutron emissions or in high velocity collisions); they are not seen without the help of these unusual events.

We have been noting how the proximity to a center relates to mass and frequency; let us expand on this a little more in the following chapter.

# 40

# PROXIMITY TO THE CORE

*T*HE *LARGE HADRON COLLIDER (with its search for, and supposed capture of, the Higgs boson) represents an expensive dead end in scientific inquiry. It is similar in scope to the giant cathedrals of old—immense and impressive in structure but of little real value beyond sustaining a priesthood.*

*Mass and frequency are equal measures of proximity to the black-hole core. They each are based on the density of expanding concentric shells, 3D shadows representing 4D reality. Our world consists of 4D Planck lengths projected and, thus, expanded with literally nothing between. One can explain, using this scenario, electron orbitals as well as deep, massive objects within protons (Z, W, and Higgs particles). Therefore, each hypersphere (hydrogen atom) is really an entire universe projected onto our 3D plane with almost all of it still hidden in the vastness of the 4D.*

## Large Hadron Collider

**Question:** Since we have been discussing particle collisions, before returning to your ideas why not mention the recent success at the Large Hadron Collider—the supposed discovery of the Higgs boson. This has been announced with great fanfare, leading to a Noble Prize for the concept's originator, Peter Higgs. But you already stated that the search is futile. Do you wish to expand on that point?

*Answer:* The Large Hadron Collider sends protons hurtling head-on into each other at speeds approaching that of light. Numerous particles are found, and I am sure that many unusual entities will be discovered. However, these are merely dormant objects shaken from the depths of the fourth dimension. They do not exist in our day-to-day world and need not be uncovered.

The whole concept of these massive colliders, many miles in circumference, is oversold. Of course they discover abundant particles, which certainly give rise to new theories, but their real value is dubious. They are essentially very expensive toys given to particle physicists, and their astronomical costs severely disrupt other basic investigations. Instead of spending billions in their construction and maintenance, the money could more rationally be used in other fruitful and essential pursuits.

They are like the cathedrals of old: giant, beautiful edifices, of great importance in sustaining a priesthood, but of little lasting scientific value. The Higgs mechanism (supposedly essential in bestowing mass on other entities) is a vestige of the Big Bang theory; both (the Higgs and its *raison d'etre*) are fundamentally flawed.

## Mass And Frequency

*Question:* So if the Higgs mechanism is a nonentity, what then allows for mass or substance in our world?

*Answer:* Mass is a byproduct of the force toward the center. It exists in our three-dimensional world as a balance (the inertia) to this attraction. The greater the force encountered, the greater the mass. Remember, centripetal movement (the pull toward an imaginary direction—$[i]F$) is equal to the energy of the universe ($E$):

$$(i)F = mc^2 = E.$$

Since $c$ (the speed of light) is a constant then the force toward the center ($[i]F$) is proportional to mass ($m$). This force is also proportional to frequency since, according to the *Planck relation* (or formula):

$$E = hf, \text{ where}$$

$h$ is a constant (the Planck constant), and $f$ is the frequency or number of cycles per second. Thus, since

$$E = hf, \text{ and, since}$$
$$E = mc^2, \text{ then}$$
$$hf = mc^2.$$

The variables (the terms in this last equation that are not constant, $f$ and $m$) must move proportionally (mathematically in the same direction). Thus, as $m$ increases or decreases, so must $f$, and vice-versa. Hence, mass ($m$) and frequency ($f$) are similar, are related; they each express the force toward the center ($[i]F$).

Remember, we described the frequency of any palpable force as secondary to the disruption of a center (the closer the more frequently felt and the greater the intensity). Thus, frequency and mass equally represent this force and relate to one another by their proximity to its source (the fourth-dimensional abyss in which a hypersphere resides). They are different ways of noting the density of energy spheres or closeness to that point.

## Closeness To Center

*Question:* Let me see if I understand what you are trying to describe. The center of each and every entity is a black hole, an area of such great attraction that nothing, not even light, can escape. This is really a higher-dimensional orb, but its shape or characteristics are unknowable to us; all we can understand is the force it generates.

At each Planck interval (the least amount of space possible), another energy sphere is reflected out to our world (the surface of a fourth-dimensional globe). Because of the distortion that occurs when a higher dimension is discerned in a lower one, an ever-increasing redshift is obtained. These Planck intervals—energy spheres—comprise our real world. Thus, the farther from any center, the greater is the width of adjacent spheres and the less dense or massive is the object.

*Answer:* I think you understand. Remember, the rotation in three dimensions is what allows for tangible substance, for centrifugal expansion to counter centripetal pull. This rotation becomes the proton surface, and the electron cloud, our basic entity, our hydrogen atom.

But this rotation can be disturbed; the force tugging inwardly can then be more easily transmitted and felt farther from its source as these clouds of substance are broken. The opening, or entrance into an individual cloud as it circles the black hole, gives rise to the frequency of the conveyed force.

The closer that cloud is to the center, the more *frequently* that opening passes any point; thus, the closer, the higher is the frequency. But the closer, the greater is the force; therefore, the higher the frequency, the greater the force. At the same time, the closer, the more dense, are the Planck intervals; hence, the greater the mass. Thus, the mass and frequency are but two aspects of the same condition; they are both signs of the intensity of the force of attraction.

## Photoelectric Effect

*Question:* I understand what you are trying to say: Frequency and mass are really similar. They are both due to the proximity to a core. However, in the photoelectric effect, first elucidated by Albert Einstein, the ability of light to dislodge an electron is solely due to its frequency (or color); it has nothing to do with its luminosity. What then does intensity (or brightness) measure?

*Answer:* The intensity is simply the size of the opening in that covering material (the electron cloud); the greater the entrance, the greater the disruption, the brighter (or more luminous) the light. But brightness of a low frequency will not dislodge electrons; only the higher frequencies will. Thus, brightness tells us *how much* disruption has occurred, but frequency is a sign of *how strong* that disturbance is—its power or closeness to that all pervasive core.

Einstein's photoelectric theory, with high-energy light quanta hitting and loosening electrons, really is tied up with the frequency or proximity to the center from which the electromagnetic wave originates. If of high enough frequency, the energy imparted will disrupt the cloud, causing an electron flow or current. But it is not a quantum of energy (a massless photon) traveling at $c$ that interacts and causes dislodgement. It is simply a wave of energy that moves through a sea of surrounding particles (photons present about all atoms) and, if of great enough frequency, causes a change.

## Quantum Jumps—Orbitals

*Question:* I seem to grasp your ideas. The space surrounding a proton consists of concentric shells, lower-dimensional reflections of a hypersphere. When tangible, we call them electrons, and they must move in distinct orbits—in energy spheres—since that is all that can exist. When impacted, these orbits can be disrupted, and this disturbance is our electric current.

*Answer:* Yes, that is what I am trying to convey. These paths in which an electron can travel are ever-widening orbs redshifted to us from a higher-dimensional sphere. Between them there is nothing—fourth-dimensional space. Thus, the electron, tangible energy or the counterforce of existence, can only be found in discrete orbits. When impacted, if at great enough force or frequency, these orbits can be induced to *jump* to higher trajectories (farther away from the center). They have acquired extra energy, pushed upward and outward; they have had their horizons expanded.

But this motion is from three-dimensional space to three-dimensional space, across a fourth-dimensional divide or nothingness. This jump must always be in discrete steps. It is not smooth, as three-dimensional space is but a reflection of Planck-length distances distorted by the redshifting between dimensions. The whole concept of energy quanta leading to orbital changes, thus, is due to the discontinuity of our world. It is irregular, it is granular; it starts and stops with ever-widening energy spheres. It becomes perceptible when a counterforce exists—when there is charge.

*Question:* So the Bohr orbitals, the surrounding electron shells, are merely how our three-dimensional world forms as a shadow of fourth-dimensional existence? The Compton wavelength, then, is simply the smallest size that the electron cloud can take if freed from an orbital?

*Answer:* You seem to understand. The actual electron is a nebulous thing. It is the energy that becomes tangible, filling spheres at ever-expanding redshifted distances. Its real size is uncertain, but the extent it attains when freed of an orbital is about $10^{-13}$ m (Compton wavelength). This is its minimal size as a palpable entity—it defines electromagnetic force. From this range it can expand, filling many distant spheres until finally so diffuse it no longer is a discernible covering for some nuclear core.

## Nuclear Shells

*Question:* If the electron is just congealed energy filling concentric globes of three-dimensional space, what about the many, many comparable but much tighter shells within a proton?

*Answer:* The inner aspects of any nucleus should present similarly; only the stacking of spheres or shells would be, as you have noted, much more compact. Therefore, the great density of matter (the ever-impinging orbs) as we approach the Planck interval leads to more and more massive objects. But they are similar in all respects to our diffuse electron clouds.

The proton and neutron are the only stable entities; but smaller, tighter wound objects can form during high-energy collisions: The tau particle, with a heft almost two times that of a proton, is such an example. When found (when fleetingly residing in our three-dimensional world), its mass is discernible. However, in short order it disappears (reenters the fourth-dimensional void) and its presence (its mass) no longer registers.

Therefore, only three-dimensional representations of these innumerable (8 x $10^{60}$) energy spheres present as mass in our universe. But the total mass of any hypersphere, if conceivably existent at one time, equals that of our world. Thus, the energy potential of the boundless quantum vacuum (the fourth-dimensional void) is much, much greater (about $10^{120}$ times) than noted; it is equal to all of the hyperspheres (real, potential, or virtual) contained in the total volume of our universe.

## Hidden Mass And Vacuum Energy

*Question:* But I still have a problem with understanding how a proton, or anything of limited size and mass, can contain within it innumerable objects of much greater heft. Would not these other entities register somehow in the real world? How can something seemingly not that weighty contain within its core heavier objects?

*Answer:* When the proton is sent scurrying about in a particle accelerator, it achieves a velocity close to that of light. Thus, according to Einstein's theory of relativity, its mass increases, and upon impact pieces of it can be quite weighty (the Higgs particle is over 100 times as bulky as the initial proton itself yet originally contained within it). Thus, mass increases with velocity, and this is sensed upon collision.

To really understand this, however, we must picture the higher dimension. Remember, the proton is the portal to a fourth-dimensional hypersphere. Behind it resides most, almost all, of that entity. Thus, within each proton (or entrance) there are innumerable ever-tighter wound orbs ending at a black-hole center, each of greater heft than the proton. However, all we can see or understand is the entrance to these many globes—the surface that exists in our world, the proton.

When hurled about and smashed into other protons, these denser orbs are momentarily freed and sensed fleetingly entering and leaving our plane. Thus, one finds a whole zoo of exotic particles (the W, the Z, and the Higgs boson, to name the most significant ones—each close to 100 times as massive as our proton) surviving but for the slightest instant. Perhaps an attempt at drawing this would help to explain the concept (see next page).

Therefore, when we find a proton we sense a doorway to a higher

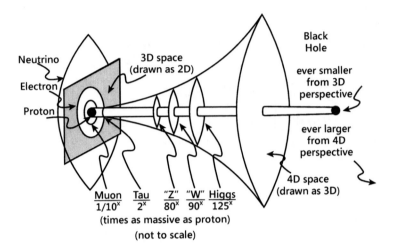

Black Hole

ever smaller
from 3D
perspective

ever larger
from 4D
perspective

4D space
(drawn as 3D)

3D space
(drawn as 2D)

Neutrino

Electron

Proton

Muon  Tau  "Z"  "W"  Higgs
$1/10^x$  $2^x$  $80^x$  $90^x$  $125^x$
(times as massive as proton)
(not to scale)

level. Within are more and more massive entities; however, they are only measurable if, due to some tremendous collision (particle accelerator or cosmic ray bombardment), they are spewed into our world. But they truly exist and are accounted for in the inordinate vacuum energy content of fourth-dimensional space.

Remember, Einstein equated mass with energy ($E = mc^2$). Mass, then, is simply another way of discerning the surface tension (energy) that must exist to counter the force toward an ever-present black-hole center. The proton is the cover (the entrance to the other plane), and its mass (its inertial energy) is what maintains its three-dimensional existence. However, within it is a higher dimension with all the mass (or energy) of our universe, which is occasionally discovered following some great impact. Therefore, mass only exists in our (three-dimensional) world, as our world is the reactive force, the inertia, to a higher realm, and matter, or substance, is its definition.

*Question:* I seem to understand. The hypersphere is really made up of many, many separate orbs, but only the proton and its electron cloud are tangible in our world. Most of the other similar entities reside in the fourth dimension, in the abyss. They can fleetingly enter but must quickly return. However, they are all real; they all have mass. When sensed in our world, for that split second, their presence, their heft, registers. However, they quickly depart to reside on the *other* side of reality. Since they all actually exist, their total mass or energy is much, much greater than what we sense in our tangible world (over $10^{120}$ times as great).

*Answer:* I could not have stated it better; you truly do understand. Why not try to summarize some of my concepts, perhaps making them simpler and clearer?

## Restate

*Question:* Okay, let me attempt it. Our universe is a three-dimensional surface, the facade of a fourth-dimensional hypersphere that travels inexorably in an unknowable direction; we consider this movement time. We are but one of many, many such entities. We visualize the others as hydrogen atoms (proton cores with electron covers). However, each one is a separate world, and each encompasses all.

Due to this rotation, we sense a force, gravity. It is the same as the strong, electromagnetic, or weak interactions; the others just appear much greater, since they are that much closer to the source. All are due to the centripetal pull that is the basis of the energy of our world.

Since we are three-dimensional, we visualize fourth-dimensional objects as ever-enlarging concentric spheres, with anti-spheres, limited by the edges of our existence. The redshifts are simply how our dimension understands and allows for a higher plane. The ever-increasing widths of these orbs, as one leaves the center, causes a rarefication, or thinning out, a decrease in density or mass. Reality in our world is the counter to the continual inward pull; it is the inertia of existence.

Frequency and mass are measures of the universal force, and Einstein's photoelectric effect is how waves of energy (seen as they traverse the photonic seas) impact electronic shores, dislodging particles (electric current) when and if these waves are of great enough strength (frequency).

*Answer:* You really do understand. Once we use this concept of rotating fourth-dimensional spheres, we can do away with a lot of accumulated baggage. The Standard Model, with its Higgs mechanism, giving particles the mass they already have, is no longer needed. The paucity of antimatter is found to make sense. Photons, once given substance, become the *light* of dark matter.

Thus, once this paradigm is grasped—rotating higher-dimensional entities, spheres with anti-spheres, each but one Planck length apart (the smallest allowable interval) yet presenting, when reflected to our world, as ever-widening concentric orbs—the other concepts are greatly simplified. Layer upon layer of theory no longer required can be discarded. There can be a great cleansing.

*Question:* But don't you have any qualms about the validity of your beliefs? Don't you worry about whether the structures you have built are real?

*Answer:* Of course I am concerned about them, because they are so radically different. Let me try to analyze some of my ideas. Then we can see how they affect us in our day-to-day existence.

# 41

# RE-EVALUATE

M ATHEMATICS SHOULD EXPLAIN THE REAL WORLD, not be abstruse for its own pur-
veyor's pleasure. Thus, the math used is relatively simple.

A guide is suggested and used throughout—essentially, the 4D perceived as
a universal force, as God.With this guide, an attempt is made to explain the diffuse
workings of our world. Its neatness hopefully is not its undoing.

## The Language Of Math

Let me start with the math I have used. It is relatively simple; most
scientists use much more complicated tools. However, to me math is
only a language to explain what exists, and it should therefore help us
understand the real world. If mathematicians want to pursue it for abstruse pleasure, that is their choice, but to me math should be useful and
explanatory.

Quantum mechanics has become so complex, so difficult, some fol-
lowers would say that, if you can explain it, you don't understand it.
Ptolemaic math was equally obscure. The math used to explain dark energy is also quite complicated. But some of these concepts turn out to
be little more than diffuse puffery and have no meaning in the real world.
Math should be a guide and, therefore, not overly difficult to follow.

## Aristotle, Newton, And Einstein

The great theorists—Aristotle, Newton, and Einstein—explained how the universe worked based on the accepted knowledge of their time. Aristotle's ideas of the universe, the foundation of the Ptolemaic system, held sway until the time of Copernicus, for almost 2000 years. Newton then explained the concept of gravity with the aid of his newly discovered calculus and infinitesimals. Einstein further refined the concept of gravity by using a *bend* in an unknown direction to describe it.

I am trying to clarify these concepts. There are no infinitely small lengths, no infinitesimals—the smallest distance the real world allows is the Planck length. Einstein was right; there is a bend—it is real and into the fourth dimension. We see it as we peer out into deep space.

## Universal Guide

I propose guideposts. The use of the fourth dimension allows for a simpler explanation of the universe, since it does away with early rapid inflation and dark energy. It helps one to understand quantum mechanics; it offers a different explanation for electromagnetic force. Some of what I propose is based on real evidence; some is deduced from the fourth-dimensional concept.

The process is similar to that of sailing a boat. Given a guide—a compass or the heavenly vault of stars—changing currents and treacherous winds will interfere only slightly with one's voyage. One can still reach the appropriate destination. The same is true of our understanding of the universe. Given God, the universal force that permeates all and is the basis of all, one has the pilot that allows for a successful journey.

## Does It Make Sense?

*Question:* So you are tying up all the basic forces of nature into one neat bow. They are all the same, they are just descriptions of that universal force caused by the rotation and geometry of higher-dimensional orbs visualized in a lower dimension. Our universe is the three-dimensional surface of one in an unknown bulk of many, many like spheres. Their presence is noted by the multiple black holes scattered about, but each black hole acts like our universe, and each has a three-dimensional exterior that interacts with all others. Thus, each universe is a part of ours, as we are a part of all.

I like the neatness of your ideas, but being neat does not make them correct. Ptolemaic theory, with its perfect circles and epicycles, was also elegant. Why aren't your theories just the crackpot musings of some dreamer?

*Answer:* You are asking me a very basic question. I often also wonder whether these ideas are too tidy, too much the figments of some foolish imagination.

However, in my own defense I must explain that they are based on solidly observed fact. The ever-increasing redshifts exist. An explanation that uses a fourth-dimensional sphere is simpler than positing an expanding universe. There is no need for imaginary dark energy or inflation. Cosmic microwave background radiation can be more easily explained.

Of course, if one accepts the conventional view, then all my ideas are wrong. But current theory needs to fantasize about 95 percent of what exists. Certainly this is a significant problem with our present understanding.

## The Essence

When one looks about, one sees that things make sense when explained by a single unifying force. This has permeated our spiritual life. Religions today are usually monotheistic. Most people fundamentally feel there is but one force, one God.

*Question:* Let me ask you once again, are you stating that God and this universal force are the same?

*Answer:* Yes. It is how God is perceived in our world. God is greater as God is the entirety—the higher dimension in all its grandeur. However, we comprehend God by the actions and forces that occur.

Let us now go from our current conceits to try and explain how a single universal force makes sense. Let us quickly travel through chemistry, to biology and life, and finally to society's interactions. Then, perhaps, my concepts may not appear to be so harebrained after all.

*Part Four*

# THE SECONDARY EFFECT: THE HUMAN REALM

T HIS FINAL PART DISCUSSES THE CONSEQUENCES *or effects seen in our world, the human realm, from an all-encompassing underlying force. We first discuss the chemistry of carbon, leading to life and evolution. Mathematics, the logic of the universe, is shown to be essential to our understanding and an important attribute that has allowed for our ascendancy—or place in the overall scheme. Religion and logic are felt to be indispensable to our true understanding, and other concepts—happiness, depression, and addiction—are evaluated. We then consider the idea of free will and how it really is a rationalization of what has already occurred.*

*In summary, this force is perceived in, and is the cause of, all human actions—an underlying essence—seen as a higher dimension in the cosmos, as an ever-increasing redshift, and the cause of all that occurs and all that exists. It is the scientific basis of our belief in an omnipotent and omnipresent God.*

*In concluding, we point toward a hopeful future based on this underlying force, and strongly maintain that true fulfillment can only come with a fundamental belief in something "greater" than ourselves.*

# SECTION X
# Chemistry And Biology

HE FIRST CHAPTER STARTS WITH CHEMISTRY, *the study of "tangible" things; it discusses how entities combine. The hydrogen atom is the basis of all other elements. Under the immense pressures found in stars and supernovas, it recombines into all that exists. Carbon is the most intriguing element, as it is capable of asymmetric formations that, by their size and shape, become extremely efficient in grabbing and holding, i.e., acquiring, energy. Under the universal force, gravity, carbon finally evolves into life forms.*

*The next chapter describes how life continues to evolve under the influence of this universal force; we are the current outcome. The second law of thermodynamics stipulates, however, that things must, under the continual increase of entropy, become more, not less, disordered. Therefore, life and evolution appear to contradict this basic law. The answer is that life, by ever advancing and accumulating, is simply the most efficient means of wreaking havoc about it, of increasing entropy. Thus, life, with evolution and disorder, are two sides of the same coin.*

# 42

# CARBON AND LIFE

C HEMISTRY IS DESCRIBED AS PRACTICAL PHYSICS. *It explains how the basic building block of nature, the hydrogen atom, combines into all that exists. The cauldron of the stars, in conjunction with explosive supernovas, ultimately create all the elements from plain hydrogen.* $CH_4$, $NH_3$, *and* $OH_2$—*methane, ammonia, and water—are some of these elements' simplest combinations.*

*Carbon, as it has four attachment sites, is capable of asymmetric combinations with itself. These complex chains, always searching for energy, finally become so sophisticated as to be self-reproducing—they become alive. Life, then, is the necessary consequence of a central pull, a search for energy. It is what the most complex element, carbon, naturally attains.*

## Tangible Things

The proton with its electron cloud, the three-dimensional reflection of our ubiquitous hypersphere—the simple hydrogen atom—is the fundamental building block of nature. Given this basic entity, we can see how things are put together.

In the cauldron of the stars, hydrogen atoms are fused into helium and then, during massive supernova explosions, into all other elements. These elements combine into molecules. From molecules we build physical things. The grouping of atoms is what makes up our tangible world;

we call this chemistry.

Up to now we have been discussing physics—albeit in a very general way. Physics and cosmology, the science of the overall shape and essence of the universe, are very similar. The same kind of mind is attracted to each discipline.

Chemistry is a little different. It is more practical. Things are put together; results are achieved. It is the science of why things stick to one another. The people who pursue chemistry are a different breed from those interested in physics; they tend less to be dreamers than doers.

*Question:* I would bet that you fall into the first category—those that like the contemplation of ideas more than the completion of tasks.

*Answer:* Yes, I like imagining or dreaming about things perhaps more than doing them. But chemistry can be discussed in very broad terms. It will lead us to biology, or the complex chemistry of life, and all the so-called natural sciences that describe our world.

Chemistry is the coming together of already-solid things, making them more substantial and finally palpable. As noted, the smallest stable tangible thing is the hydrogen atom. At its center is a proton. Around this hovers a cloud we call an electron. Its overall size allows it to attach itself easily to larger atoms; it becomes what fills the crevices of the universe.

The elements of nature are found throughout space. Under the all-encompassing force of gravity, accumulative processes occur. Atoms caught in the pull toward a black hole, or fourth-dimensional core, will swirl about, allowing denser areas to form. In significantly compact regions, hydrogen will join with other elements (predominantly carbon, nitrogen, and oxygen) to form compounds (multi-atom complexes).

When hydrogen attaches to carbon, it can combine at four available spots. To nitrogen there are three potential sites. To oxygen, there are but two. These, $CH_4$, $NH_3$, and $OH_2$, or methane, ammonia, and water, become some of the most common compounds found in space.

## The Chemistry Of Carbon—Life

The most intriguing element appears to be carbon, with its four available attachments. It can associate with all the others, but more importantly, it can combine with itself. It can form into chains, carbon-to-carbon, of any length whatsoever. Carbon easily attracts and uses hydrogen to form a stable outer shell of electrons (an octet); thus, we often find a single carbon with 4 hydrogen atoms, or 2 carbons with 6 hydrogen atoms, or 3 carbons with 8 hydrogen atoms, and so on.

The other common elements, such as oxygen and nitrogen, cannot easily join unto themselves in long chains and so are forced into alliances with others. Their repetitive patterns are much simpler than carbon's; they are usually found only in union with 2 or 3 other elements and cannot vary this with long, asymmetric runs, as does carbon.

Carbon, thus, has an advantage when compared to other entities. It has the advantage of asymmetry and complexity. Therefore, in the competition with others for energy, carbon will be more efficient—will win; it will be able to grab and hold onto more things than will any of the other elements. It will steal from the others and combine that plunder with itself. It will incorporate the others. It will strive to be more and more complex. Finally, it will come *alive*.

*Question:* You have just made a very large jump indeed; you have just glibly created life! There must be a difference between plain combinations and a force that makes things come to life. It cannot be only complexity or size, for then large rocks could be alive.

*Answer:* Well, perhaps large rocks *are* alive. Their rates of metabolism would be much slower than ours, but even among manifestly living things there are great variations. A simple virus may survive for less than a day, a mighty redwood tree for 3,000 years. The difference in life span is over 1,000,000 times. If we now take a redwood and compare it to a rock, the rock may have been in existence for over 3 billion years. This is a span of 1 million times as compared to the tree, but no greater in magnitude than the tree as compared to the virus.

*Question:* But life still implies motion, assimilation, reproduction; a rock is just a dumb thing; it performs none of these functions.

*Answer:* Once more, we do not know this, since we do not live long enough to see whether these changes occur. To a virus, a tree is forever. To a tree, a tiny, unassuming virus is here and gone in too short a time to be noticed. Thus, the perspectives of time and size are both important in calling things truly alive. To get back to carbon, we are saying that it is the most efficient absorber of energy. Carbon is more capable than other elements around it of grasping and retaining things. Carbon is the best at what it does. In head-to-head competition, it wins.

Freely floating (in the diffuse realm of space) carbon, nitrogen, and oxygen all catch and hold hydrogen; but in tight quarters (on, say, the surface of the Earth), carbon seizes the hydrogen from its competitors. The losers, bereft of their precious hydrogen or energy, are now set adrift. They become our waste products and, when finally in large enough concentration, our atmosphere.

Life, then, as we understand it, is a competition for energy at a molecular level; it is what carbon does best. We see life in our own terms, but in a broader perspective it is similar to what drives all things. It is a manifestation of the universal force—the overwhelming attraction toward the center.

## Life's Beginnings

*Question:* Assuming what you say is correct, at what point do you feel that life actually starts? What, to you, are its first glimmers?

*Answer:* Probably, we could say that it begins when some complex carbon-to-carbon molecule becomes capable of quickly, efficiently, and repetitively reorganizing less complex molecules (deriving energy— electrons or hydrogen atoms). When carbon-to-carbon becomes so well organized that RNA or DNA are formed most, then, would agree that life exists, for then carbon can remake, reproduce, itself beyond merely absorbing the energy of others.

However, before these very complex molecules occur, simpler ones can remodel (thus remake) other molecules to obtain their electrons or energy. Simpler enzymes can cause rapid changes in otherwise inert compounds. They do not appear to reproduce themselves, so perhaps they are not truly alive. Yet we are finding that, although viruses or simple DNA or RNA exist and can, with the help of intact cells, make more of themselves, even non-RNA or non-DNA molecules—prions—can succeed as well if they are in the right environment.

So I am not aware of, and I doubt that anyone truly understands, the stage at which life first begins. However, I do not consider this knowledge to be crucial, for life is the inevitable consequence of carbon's efficiency. Carbon is simply better at absorbing energy than its inanimate surroundings; it will evolve.

# 43

# EVOLUTION

*VOLUTION IS SEEN AS A CONTINUING INCREASE in complexity caused by carbon's inherent asymmetry and efficiency of accumulation. It is what all things do as they seek the central force. It is the natural outgrowth of such a force.*

*However, the second law of thermodynamics mandates that, in a closed system, things must devolve—entropy or disorder must continually increase. This apparent contradiction is examined and explained; life allows for the greatest disorder about it, the most entropy, as it evolves. Hence, order comes from disorder under the aegis of a central pull.*

## Complexity

To return to chemistry, carbon, a shrewd and efficient fighter, wins the battle with its less resourceful neighbors. Nitrogen and oxygen continue to be banished to the atmosphere. Carbon combines in greater and greater complexity on the surface, reaching the stage, of early RNA or DNA or some like molecule, in which it can reproduce itself. Perhaps now life has really begun.

However, no real difference has occurred. All that has been ongoing is a continually more and more complex carbon and hydrogen interaction. It is inevitable. Complexity is inescapable. Growth and evolution cannot be stopped.

Darwin was the first to describe this. He saw it in a continual adaptation of life to its environment. All life continually, albeit, on a very lengthy time scale, evolves into more and more complex patterns. Thus, evolution is an irreversible result of the way the universe is shaped.

*Question:* But isn't there a law of thermodynamics, or something, stating that things are supposed to get less, not more complex the longer they exist?

## Entropy And Disarray

*Answer:* Yes, the second law of thermodynamics implies that, in a closed system, things progressively get more and more chaotic. It is the concept of entropy or lack of purpose. Objects over time will, just by chance, take on an increasingly muddled appearance. This is because there are many, many more ways something can be in a disordered configuration than in an ordered one.

Take a cup of coffee, add milk, and stir. Initially, one sees light and dark liquids, but as the coffee and milk begin to mix, a tan, creamy consistency appears. The longer it is stirred, the more the two distinct liquids meld together; thus, the more disordered the solution becomes, or the greater is the entropy. Therefore, entropy is the chance that something will be in disarray. Since we are talking of individual molecules (coffee and milk) and, since there are vast numbers of such molecules, the odds of disorder are much, much greater than of order.

All things in a closed system will increase their disorder over time; all things will increase their entropy. It is almost infinitely impossible for order to come from chaos. Yet life is not only possible, it is real. Life is the rearranging of a disorganized cosmos; and evolution allows for ever-increasing structure in that same world. Since life exists and evolution appears to be a fact, and since the second law of thermodynamics also is valid, what is the solution to this puzzle?

*Question:* I raised it, but I am not exactly sure how to solve it. If entropy always increases (if disorder always becomes greater) yet life exists and becomes more, not less, organized (if life evolves), perhaps it is because the more structured and coherent something becomes, the more it must diminish its neighbors—the greater is its surrounding entropy.

## Order From Disorder

*Answer:* I think your conclusion is correct. We were discussing evolution as a continual adaptation of life to its surroundings. It is a constant increase in the ability of living things—through their offspring and later generations—to more efficiently obtain energy from their environs.

The more competently energy is acquired, the more disordered is the object from which it was obtained. Thus, evolution allows for an enhanced ability to adapt—for ever-increasing order in the developing species; but it occurs by expanding the turmoil, or entropy, about it.

Efficiency of acquiring energy, carbon's capability due to its astonishing asymmetry of attachments, due to its ability to become alive, leads to greater and greater disarray in the surrounding environment. Thus, living things are no different than the inanimate objects about them; but they are more adept at causing entropy and are more likely to succeed than their inert brethren.

The second law of thermodynamics is a fact. And life is real. There is actually no contradiction. And evolution persists forever. We, and the societies in which we live, continue to evolve into more and more complex entities.

Complexity, or life, and disorder are two sides of the same coin. Life is successful because it feeds on energy, causing more entropy, disorder, than any other expression of the universal force. Life is inevitable, as is its evolution with society. There is no contradiction between increasing complexity and increasing disorder. They are both products of the same underlying force, that inexorable pull toward the center—gravity.

*Question:* You feel, then, that evolution is the side effect of a central attraction. Things under its influence evolve; they become more and more complex. Carbon is the best and most efficient in this process, and the residual outcome is increasing entropy.

## Pull Toward The Center

*Answer:* Yes, that is what I am trying to say. The central force or pull gives meaning to what exists. What evolves when we discuss life are carbon-to-carbon bonds. The end product, so far, is us. But complexity will continue. As long as the central force holds and new energy can be obtained, carbon will continue to evolve. Carbon is the most efficient at maintaining order but, as a result, one finds the greatest disorder.

Of course, evolution in any species is constrained by the surrounding milieu. Carbon-to-carbon attachments cannot expand beyond what is environmentally possible, but carbon-to-carbon bonding will always attempt to become more resourceful at obtaining energy, more efficient at causing chaos.

Therefore, the act of grasping and accumulating energy predates, and is more basic than, life. It is the awareness of a central pull. Life evolves because of carbon's complex design, its effective ability to seize and maintain this energy. But the force, the desire to grab onto and hold, *the hunger for more*, is even more fundamental than life itself.

# SECTION XI
# *Thought And Action*

HESE FINAL CHAPTERS START by discussing math and logic as the basis of our world. Religion, a belief in something grander, mirrors and helps to explain the scientific world, the higher dimension. True religion, therefore, incorporates science in its attempt to understand the workings of God.

Our actions have consequences; they can lead to happiness or depression. The same effect is seen in individuals and societies. Addictions, and the ability to break them, are discussed. Our place in evolution is seen as in flux as computers become ever more complex with the real potential of becoming "alive."

Free will is discussed and found to be a rationalization of what must occur. We are self-aware but no more in control than inanimate or lower life forms; however, the inability to have pre-knowledge allows for the charade of free will.

Finally, the future is discussed in positive terms, based on an ever-increasing abundance as we gain further knowledge of the ultimate underlying cause. A belief in that which is higher is considered essential; its basis is both scientific and spiritual.

# 44

# MATH AND LOGIC

G*ALILEO SAID THAT MATH is the language of science. Math is a form of logic; it is based on fundamental axioms, accepted but not provable, that are then, by a logical sequence, formed into usable theorems. It is essentially the exploration of cause-and-effect relationships.*

*The universe is based on a belief that is not provable—an underlying force caused by a higher dimension. This force can be visualized as gravity or, if religious, as God. The workings of the universe, however, are due to cause and effect; thus, the universe's machinations are understandable, and math is the means to this comprehension.*

## Why Math?

**Question:** You were discussing math earlier and argued that most investigators used complex mathematical concepts to explain their findings. Why should math be the language of science? Why is it so successful in explaining hidden physical laws?

**Answer:** You are asking a very basic question that has to do with the philosophy of science and math. Most scientists employ mathematics in explaining their theories, but few wonder why it works. Einstein once quipped that the "most incomprehensible thing about the universe is that it is comprehensible," and Galileo supposedly declared that the language of science is mathematics.

## Cause And Effect

When we try to understand what math really is, we find it to be a form of logic. It is fundamentally a search for the linkage of cause and effect. Most math is a logical construct built on generally accepted axioms. Remember, axioms or postulates are basic, agreed-upon truths. In Euclidean geometry, one such accepted belief is that parallel lines do not cross. Given this, and several other similar concepts, the whole edifice of geometry was erected and has remained intact for over 2000 years.

However, in the 1800s mathematicians tinkered with this fundamental idea, showing that, in curved space, parallel lines *can* intersect. A whole new geometry—non-Euclidean—was then formulated; in fact, Einstein constructed his general theory of relativity on this new geometry.

*Question:* So you are defining mathematics as logic based on accepted truths or axioms. I guess just about anything can be used as an axiom even if it makes no sense in the real world. Thus, from what you are saying, there can be many unusual types of mathematics. Why, therefore, would these weird fabrications have anything to do with the way the universe is shaped? Why, then, should math be the language of science?

## Axioms

*Answer:* Well, even though the process of math is a cause-and-effect affair, a logical extension of given truths, and even though these verities need not apply to the actual world, usually mathematicians use concepts that do coincide, even if only somewhat, to reality. Thus, a lot of math can be quite abstruse and of benefit only to those entertained by puzzles, but similarly much can be of great value.

The closer the postulates are to reality, the more the conclusions (theorems) are in sync with the world; and the real importance of math lies in the intricacies of these deductions. So, given all the varied forms of math, it is no wonder that some turn out to be quite useful in explaining new physical phenomena.

Remember, Einstein used complex non-Euclidean geometry to explain his theories. Quantum mechanics too is based on very involved mathematical concepts. Newton even devised a whole new entity—calculus—to explain his ideas. Thus, if the underlying assumptions are valid, and if the logic used is correct, then some of mathematics can, and occasionally does, lead to important physical laws and new concepts.

*Question:* But, if the importance of math is that the logic used (once given an accepted truth) is what leads to interesting and occasionally important findings, where is the connection to your belief in how the universe is designed?

## Universal Force

*Answer:* I feel that there is one underlying force, described by religions as God, seen by science as gravity, that is the basis of all. We, carbon-based entities, are the most efficient at utilizing this force. We, because of evolution, have become more and more skilled at acquiring and accumulating the abundance wrought by this force.

We are the best at increasing the entropy around us, of seizing and absorbing the energy or essence of inanimate or less competent living entities. We are the top of the food chain, and we continually strive to become yet more proficient. But this essential act (the striving to become better at attainment), which forms the basis of the evolutionary process, is due to this force working through our fundamental carbon-to-carbon bonds—through the asymmetry and complexity that only carbon allows.

We, as humans, can contemplate this activity; our mental capacities, our brains, have become superb instruments of acquisition and achievement. We can understand cause and effect and, therefore, employ relationships to further our goals. Mathematics is a distillation of these rational connections. It is how we can identify and efficiently find increasing sources of energy. After all, logic is only a concise way of evaluating cause-and-effect linkages. Logic allows for a clearer understanding of how to get what is needed, how to find what is necessary for sustenance and growth.

## Logical Universe

Therefore, it is only appropriate that our brains are logically wired. It is simply fitting that mathematics, a form of pure logic, is so basic to human understanding. The universe itself is essentially rational. There is an underlying force, a fourth-dimensional construct, that allows existence. Our three-dimensional world is merely the counterforce, the response to a universal attraction.

Energy and matter make up our world. But matter is just congealed energy, and both are simply the reaction to this overwhelming influence, to gravity, to God. Given the axiom (the accepted but unprovable truth) of this concept of force, we then get our universe. There is cause; there is effect.

Math, as it explains this logic, is what we can use to be more successful. Math, therefore, is the language of science, the language of how the universe functions. The universe is essentially logical and, to re-phrase Einstein, it should not be incomprehensible that it is comprehensible.

*Question:* So, to you, math should be the language in which science is written, for math is, in its purest form, the logic of the universe; but isn't there some fundamental hypothesis that, in essence, states that mathematical concepts cannot be proven?

## Gödel's Theorem And The Brooklyn Bridge

*Answer:* You are alluding to Gödel's theorem, which basically notes that an edifice, a theory, can only be proven correct if the axioms upon which it is constructed can be found to be true. Since axioms are always *accepted facts*, not *proven certainties*, all mathematical structures are inherently unprovable.

But math need not prove a truth; math is just a logical process. All that needs to be accurate is the process itself. If the logic is consistent, if there is real cause and effect, then the math is correct. It is not provable; it is, however, usable.

Take the example of the Brooklyn Bridge. The towers were to be anchored in bedrock for stability, but as deep as the workers dug, no underlying rock could be found. Finally, the towers were erected on sand. They have remained entrenched and intact for over 100 years. Thus, although the foundation is not as secure as possible, the resulting structure is of great use and beauty.

## The Real Beauty Of Math

*Question:* So, if I understand what you are saying, mathematics—a logical process for efficiently increasing energy—is an outgrowth of the human brain. It is an adaption caused by evolution. It is that for which all animate matter strives.

We, the most advanced of all living things, become more and more adept at cause-and-effect relationships. Thus, we have constructed math. Math can be useful, explaining facts or physical laws, or merely abstruse—engaged in for the pleasure of the seeker.

The reason it is so important, the reason it is the language of science, the reason the universe seems to be comprehensible, is that there is a logic underlying the myriad and seemingly disparate occurrences of our world. This logic is the same as of what we are composed.

*Answer:* Yes, the beauty of math is the process; it nurtures and nourishes us. Occasionally, it explains the real world, thereby accelerating evolution—making us more alive.

# 45

# TRUE RELIGION

F UNDAMENTAL RELIGIONS OFTEN BASE their ideas on the literal interpretation of a holy scripture. Everything that has occurred is to be found in these writings; there is no concept of cause and effect. All is due directly to God; there are no intermediary steps.

"True" religion, however, takes into account the underlying essence of a higher dimension but also allows us to see how its effect is established. It uses the same logic as does science, the equivalent of math. True religion allows for continued growth and evolution, enabling us to contemplate the workings of God.

Dreams are a process; their content is superfluous. They allow us to discard the day-to-day meanderings that occur and to focus on the essentials: They are in a sense the voice of God.

## Platonism

**Question:** I like your previous idea of math being the logic of cause and effect, but at times your concept sounds too simplistic, too biased. Isn't there an entire school of thought that considers mathematics to have an idealized, almost Platonic existence?

**Answer:** Yes, many mathematicians consider math to exist by itself in a perfect form, one that has a basis in Plato's description of the world.

He felt that all we can see or know is but a vague replica of reality, that the true essence of substance exists as a pure concept but can only be sensed by its reflection or shadow.

Although we have already discussed this (fourth-dimensional spheres visualized as particles and waves), whole, macro constructs do not exist in the micro world. There are no finished entities in a distant, higher dimension corresponding to those in our three-dimensional world. Our world forms in its three dimensions by cause and effect, by force constantly tugging toward centers. It is not a replica of discrete, idealized higher-dimensional forms. Platonism, then, would appear similar to many fundamental faiths where everything is possible, as all is but the will of God.

## Religious Fundamentalism

In many fundamentalist beliefs, be they Christian, Jewish, or Muslim, or even in other, non-monotheistic religions, one literally adheres to some revered book. The teachings, taken word for word, often go against everyday reality.

Take creationism (a literal interpretation of the Bible), which views evolution as an evil perpetrated by a sinister, ungodly science. It is based on God being omnipotent—all creations are God's alone. Anything willed by God is possible; anything written in the Book is literally correct. Creationism denies cause and effect. All is initiated directly by God. There are no intermediary steps. There is no trail of logic, no math.

Or consider the Catholic Church at the time of Galileo. The real reason Galileo had to recant was not that the Church felt his ideas were necessarily *scientifically* wrong; it rather felt that *any* explanation other than God's will was needless and incorrect.

Or think about fundamental Muslim belief today. In many of the teachings, whatever occurs is directly due to God's desire. There are no intermediary steps. There is no cause and effect. Anything becomes possible, as all things are willed and directly controlled by God.

## Religion's True Meaning

*Question:* So to you, literal readings or idealized concepts are similar. To you a proper understanding means a visualization of a universal force, but also a cause-and-effect relationship. It means the force works via logic, via math.

*Answer:* Yes, that is how I see it. True understanding is of God, universal force, and existence via logic or cause and effect. God still is the reason for all, but God is not capricious. God does not, as Einstein noted, play dice. God has caused and continues to cause all the effects, but the process, the logic, the math, is decipherable by his subjects. God is mysterious, a higher dimension that cannot be comprehended, but the effect of God's existence is felt and can be understood.

I feel that true religion takes into account the striving to understand, a desire that is essential to all. It incorporates science. It utilizes the language of math. It is the logic of the universe.

*Question:* But did not Gödel show that logic is uncertain, that we cannot prove our assumptions?

*Answer:* You are correct; however, logic is simply a process, a mechanism that allows for greater and greater efficiency and life. So true religion calls for, it demands, an understanding of what exists; it allows us to consistently strive to help realize the essence of all, to contemplate the workings of God. True religion is *a part of*, not *apart from,* this world.

## Why Dreams?

*Question:* What you are discussing is society's effort to acquire and accomplish. But if logic, or math, or science, is what we understand while consciously achieving this goal, what about our thoughts when we are asleep, when our awareness is altered?

Remember, you alluded before to being more of a dreamer than a doer. Dreams are very strange. Most animals appear to dream, and we spend about one-third of our time in such a state. There has to be something essential to dreams, as we are completely defenseless while they occur.

*Answer:* People have always attempted to interpret dreams, but to try and understand them by what they represent is the wrong approach. They are aberrations of occurrences in our mundane existence. If we look into them, we find our encounters and troubles of previous days.

They serve, however, a more basic purpose than just a rehash of our daily pursuits and difficulties. They are how the very force of the universe is interpreted and reviewed. They allow us to separate the *wheat* from the *chaff.* They allow us to get rid of the unnecessary burden of day-to-day existence, the extraneous, unneeded facts that we have accumulated, and get to the essence of the underlying force.

We, animate carbon entities, are much more efficient than other things in acquiring and accumulating energy. We are always seeking the source, and dreams allow us to hone in, to become more proficient at attainment.

In a sense, dreams are the voice of God. They are our minds' adaptation to effectively seeking and using the energy of the universe. That is why most animals willingly become defenseless, easy prey, for a substantial portion of their lives. The trade-off is a greater efficiency in attaining essentials. It is an offset that leads to enhanced growth and future success.

*Question:* I guess what you say makes sense. Since most animals sleep and appear to dream, the necessity to dream is practically universal. We, as animals, dream, but we, as intelligent beings, attempt to interpret these dreams. However, since it is very unlikely that any other animal analyzes its dreams, yet most sleep, the act of investigation appears unnecessary.

*Answer:* Yes. Thus, the classic interpretation of dreams is of little value. The importance of dreams is found in the act of dreaming; their weird meanderings do not require evaluation. It is the process, not the content, that is essential.

# 46

# ACTION AND CONSEQUENCES

*T*HE UNDERLYING FORCE OF THE UNIVERSE, *gravity, is seen in all that occurs. It is the force behind our need to acquire and accumulate; it is what makes for ever more efficient societies. The inability to maximize this force can lead in human terms to unhappiness and in society's terms to bankruptcy and depression. The pursuit of this essence is our unalienable right—the pursuit of happiness.*

*Time fillers such as excessive work, games, and drugs all have an addictive effect that is very hard to overcome. Probably the most difficult "addiction" to break is overeating, as it is a direct consequence of how the underlying force of the universe is experienced. Religion is a spiritual, not a physical, pursuit, but its hold on many of us is very powerful.*

*As our societies continue to evolve, newer and more efficient means of acquisition are sought. Computers continue to exponentially improve and may in the foreseeable future become as intelligent as we are. Once so complex, they will seek energy for themselves as much as for us. We, therefore, are at the cusp of a very exciting time—organic, carbon life forms morphing into higher, more complex and efficient "life" forms.*

## Depression—Individual And Economic

**Question:** You continually talk of an underlying force as the basis of everything that exists; therefore, must it not also be the source of all human activity?

*Answer:* Yes, this force that pervades all things, that makes carbon more and more complex, that forms life, also underlies all of our actions. Our desires to accumulate, to accomplish, are manifestations of this essence. We are more in tune with the natural order if effective at these tasks. Thus, we feel more content or happier. In a similar fashion, when unable to acquire, when out of sync, we feel less satisfied: We are unhappy or depressed.

Society is the totality of all individuals' strivings, and economic problems occur when these actions are thwarted. Since there is always a force of accumulation, in capitalist societies groups (such as banks) lend money to achieve this need. Even though the value of the money lent remains a constant, the desires and abilities of the borrowers vary greatly. In general, the best loans are given first, and then loans of lesser merit are distributed, until, finally, the poorest-quality ones are allowed to those who wish to succeed but have no genuine plan of how to do so.

Since the money lent is always of uniform worth, but the debtors' capabilities differ widely, the poorest loans are frequently not repaid, and society can then become bankrupt (it may enter an economic downturn, usually a recession but occasionally a depression). Thus, the underlying force of the universe, seen in human terms, as a need to accumulate and accomplish, can and occasionally does end in unhappiness and depression.

## The Pursuit Of Happiness

*Question:* If we perceive this force as a need in an individual to attempt to acquire (to amass a fortune), should we then allow actions that might stifle it? Does it make sense to diminish this drive with excessive inheritance (the bestowing of extreme wealth), given that one's happiness is so dependent on individual determination and striving?

*Answer:* This is an intriguing question. As the need to *attempt* to succeed is fundamental, and the thwarting of this can lead to depression and unhappiness, is it appropriate to pass great amounts of inherited wealth to children? Probably it is not, as we are now blocking, in our offspring, the necessity for this pursuit. By inordinate gifting, we decrease the desire to strive—to do one's utmost. We diminish that essential *pursuit of happiness* to which we are all entitled.

**Question:** Well, if excessive wealth is really a hindrance, as it frustrates one's need to try, does not excessive welfare do the same?

**Answer:** Yes, excessive welfare, or inheritance, are both inappropriate, as each decreases the motivation that makes for one's happiness. They both diminish the individual to whom they are given and are not advantages, as some may think, but actually detriments.

## Breaking Addictions

**Question:** Earlier we were talking about heart and vascular problems as secondary to over-eating, over-accumulation. Many of us find it extremely hard to lose weight. Is this because not eating essentially goes against the very fabric of the universe?

**Answer:** Dieting, or weight loss, is very challenging for most because it really does subvert the essence of the universe, the very way we perceive force. The desire to eat, to acquire and accumulate energy, the hunger for more, is the way this force manifests itself in an individual.

**Question:** What about the other distractions, the other ways of filling civilization's empty time: alcohol, drugs, games, excessive work, religion, gambling? Do they also share in this force?

**Answer:** The other so-called addictions or pursuits are really secondary effects, not directly caused by the universal attraction. Thus, gambling imitates the actual force of accumulation, whereas alcohol and drugs mimic its derivative, the sensation it gives when attained. Religion is of an entirely different order—it allows meaning to this force.

If we were to rank the importance or strength of our needs by the difficulty in blocking any individual one, then we would have to make over-eating the most significant, followed closely by gambling, work addiction and games, then drugs and alcohol. Religion, as it is an intellectual and spiritual pursuit, cannot be grouped with the more physical desires, but obviously has an exceptional hold on many of us.

## Computers—A New Life Form

*Question:* We previously were discussing evolution as an expression of the universal force that, through carbon's asymmetry and complexity, has led to all the intricacies of life and society. Why don't we further explore it now?

*Answer:* That is an excellent idea. We appear at the present time to be on the cusp of an evolutionary change. We are going from carbon-based to non-carbon, or inorganic life forms.

Implements that we use get imbued with life. A tool, a spear for example, becomes an appendage like a hand to an arm. If used routinely to acquire energy, that device becomes animate, part of us. It obviously does not live by itself but does so with the help of those who make and use it. A knife is neither more nor less than teeth or sharp claws; a knife is an object made for us, whereas teeth and claws develop with us, but they are the same. They evolve for the more efficient use and acquisition of energy.

Today, although we have many different instruments, the most important, by far, is the computer. It is the first device that approximates our intelligence. It is becoming more and more sophisticated, and in the next fifty to a hundred years should become as complex as we ourselves, and after that (in some of our own lifetimes) it will surpass us.

We will still build it, but it will evolve beyond us. Its goal will always be the acquisition and incorporation of energy. We should, at least initially, be the benefactor, but it will become to us as we are to a DNA molecule. We exist because DNA reproduces itself, but we are much more than a complex molecule. Computers exist because we design and build them, but they will become vastly superior to simple humans. Thus, evolution is forever. We are at a watershed time, and I for one find it very exciting.

*Question:* You may be intrigued, but you almost make it sound like the movie *The Terminator,* in which, as soon as computers become self-aware, they become evil. Aren't you afraid this could happen in real life—that once computers become so advanced, they may wish our destruction?

## Singularity

*Answer:* You are voicing the fears of those who feel there will be an abrupt change, a singularity, once the computer becomes self-aware. They liken this to the supposed singularity of the Big Bang. But just as there was never a sudden beginning, there, too, should not be an abrupt change from human to machine intelligence.

What most likely will occur is a slow evolutionary process in which more and more information will be maintained by machines until their self-awareness becomes obvious. However, initially, we will still be essential for their support. Slowly, over the years, we will become merely useful, then, finally, of no intrinsic importance. There should not be a singular break but more probably a slow divergence of interests.

Besides, the computer's increasing ability is a fact. It cannot be altered by anything we do. Computers will evolve and become ever more competent and, in the not too distant future, transcend us. Competition, society to society, will be the driving force. Fierce rivalries will force the most technologically advanced to constantly improve their machines; for if one group does not progress (or keep up), it will be bettered—it will fail. A computer, although the most advanced device we have, is still currently but a tool. Its designer's objectives remain, as always, the incorporation of energy. Thus, it will be continually honed into a better and sharper instrument. We cannot stop this evolution.

Whether these changes will be good for us is an entirely different concern. In general, the more intelligence we gather, the better becomes the lot for the great majority; there is more food, less illness, and a greater longevity today than ever before. People may or may not be happier, but the fact is that we are physically better off than in prior years.

I, for one, do not feel that more knowledge in the community is harmful; it will enable us to better organize and enhance our food and energy needs and our social amenities. I also feel that the computer's best interests will, for the foreseeable future, be the same as ours; for, although it may become more adept, it will still require our maintenance for its growth. However, it is too difficult to predict when, or if, in the distant future, the computer's desires may significantly diverge from ours. Besides, there is little point in trying to stop the inevitable, lest there be a catastrophe of Biblical proportions the computer will evolve from, and surpass us.

## Genetic Engineering

*Question:* So you choose not to worry, or at least not to think, about the side effects, for they are inevitable. However, we are not just the maker of the computer, an inorganic life form; we are also the designer of genetic changes in bacteria and plants. Aren't these also potentially very fast-growing and complex sides to our evolution?

*Answer:* Yes, genetic engineering is indeed a new and very important tool in our culture. It will hopefully make our life better. It, like the computer, is a vast new opportunity to remake our world. But unlike the computer, it is not yet, at least, a copy of our most human trait, our mind. Genetic engineering will improve our lot, but computer evolution will change life.

# 47

# FREE WILL

A CTIONS HAVE CONSEQUENCES; CAUSES HAVE EFFECTS. *The overall design of the universe is such that every object is attracted to every other one. All things are pulled toward the "center," by gravity, by God.*

*This force permeates all entities both animate and inanimate. We have as little control over it as does a comet hurtling toward the Sun. We are always in the present instant, but it is too short to be cognizant of; thus, all of one's thoughts are of the past.*

*As thinking, sentient beings we are self-aware. We see what has been preordained and rationalize it as if it were our own choice. Time to us inexorably moves from the past through the present to the future. From a 4D perspective, the universe consists of innumerable 3D worlds aligned one after another; therefore, from a 4D viewpoint, there is no time. Thus, the future is already set; however, to know it we must live it. As we cannot have pre-knowledge this lack allows for the charade of free will.*

## Toward The Abyss

**Question:** You speak of human actions and their consequences. You always claim that we are trying to maximize energy. You stress that we, carbon entities, are drawn by that higher-dimensional attraction toward the source of all, that unknowable region, to God. This pull is inherent in the very way we are composed. We cannot resist it; it is of what we are.

You also state that we travel toward that dimension thinking we are moving in time. We travel inexorably to the future. It keeps expanding in front of us. We continuously enter it. We obviously have no choice, we cannot stop time—we cannot stop the universe from moving forward into the fourth-dimensional abyss.

But what about free will? Many religions demand it of us; we must be endowed with it, we must be able to make choices, must we not?

## Moment To Moment

*Answer:* Your question is very difficult to answer. If all of our actions are an effort to maximize energy, and if all of them are due to a universal pull felt in the very fiber of our being, in the bonds that hold carbon together, do we really have any control in what we do? I think our concept of free will, of making choices, is a charade, a rationalization of what occurs.

We are in a constant flux, always in the present instant, remembering the past and hurtling into the unknown, the future. But this present instant is fleeting; it lasts just one Planck moment. In one second there are almost 50 million, trillion, trillion, trillion such instants. Each is real, but each is obviously too short to be felt.

The infinitesimal changes instant to instant constitute motion in our world. Each moment is a frozen frame slightly altered one to another; these constant alterations are the essence of how the world works. Each frame is a three-dimensional image of our universe stacked one after another toward the unknown, the fourth dimension. If we were a higher-dimensional *being*, we would comprehend all these individual instants; we would sense time unrolled from the past through the present and into the future.

## Absence Of Time

Thus, for such a being (if one existed), there really is no time. Each instant has already occurred; all that has or will happen is currently present, together at one moment. Therefore, all that we have done (all that the universe has accomplished), and all that we will undertake, already exists.

To us, the future, the distance into the unknowable fourth dimension, cannot be known. We must live it to reach and experience it. We cannot have knowledge of what will occur, we can merely infer from what has passed what likely will happen; but it is always a guess based on the overall inertia of the universe.

## Rationalization

Therefore, if the future already exists, if our actions are already preordained, where is our alleged free will? In reality, it does not exist. It is simply the rationalization of an action that has already occurred.

We are sentient beings; we think about our surroundings and our actions. We feel that we are in control and can do what we wish. But in actuality, as each instant brings slight changes, and as there are so many instants, so many changes of which we are unaware, we truly are not in conscious control.

We are no freer than a comet hurtling toward the Sun, just missing, then flung out into space, only to orbit again and again. That comet is not sentient, it does not think, it is not aware of what it does; it obviously does not have free will. But, if it were a thinking being, would it not also consider its journey of its own volition. Would not that comet also say to itself, "Today I am getting closer and closer to that bright spot and by choice will circle about it and fly away"?

## Animate, Inanimate, And Choice

**Question:** But how can you equate conscious beings with inanimate things? We are alive; we think and desire, we move and act. We control our own destinies; we must have free will.

**Answer:** Earlier, we were discussing the difference between living and inanimate objects; the difference is not fundamental, it really is one of degree. Living and nonliving entities all are part of ever-increasing entropy, of increasing disarray; only living ones are much more efficient in causing this disorder about them. That is what life allows. Making choices allows for those alive to more ably, and in greater quantity, acquire energy than inanimate ones. But all things pull all others toward themselves—all objects exhibit gravity.

A flower grows toward the light; a cat crouches and springs on a mouse. These are both under the control of that universal force. A flower, if it could think, would marvel at that wonderful source of energy and try to reach it. A cat is self-aware but does not contemplate its own movements; it simply acts. Both are alive, but, just as the inanimate comet, both are controlled and predestined in their journeys.

Because living things are so efficient, there is great disarray about them. Animate beings make choices, attack and destroy, acquire energy, all very effectively. But the underlying force is the same: the central attraction, the force set up by our three-dimensional surface as it glides over a fourth-dimensional abyss—the universal centripetal pull of gravity.

## Cause And Effect

Humans are conscious of their own and other beings' actions. We give a rational basis to what occurs. We devise math, the essential logic of cause and effect; we find subtle differences and exploit them. We continuously adapt to our surroundings; we constantly acquire. Our societies grow ever more sophisticated. We obtain and store knowledge; we speak and write; we record and remember.

We, human beings, ponder what we do and, hence, consider our planning and effort as its cause; but really the same force that attracts an inanimate comet underlies our actions. That same essence controls all that we do or contemplate doing.

## Fleeting Instants

*Question:* So if the present instant is too short to be known (if we can only be aware of the past), then all that we are cognizant of must have already occurred. Although everything is constantly happening in our fleeting present, we cannot know about this moment until it has already passed.

*Answer:* Yes, that is what I am trying to say. All our thoughts are of the receding past. The process of thought is much, much slower than the actual changes that almost instantaneously occur. These changes are due to the force that pulls us toward the center, the unknown fourth dimension.

All our thoughts, all our supposed reasons for action, are rationalizations of what we have already done. We move, we cause effects, we have desires—we continuously enter the future. But our thoughts are not the real reasons for these actions; our thoughts simply attest to their existence.

## The Unknowable Fourth Dimension

*Question:* So the future is unknowable; it is the fourth dimension. We must travel toward it constantly. We perform myriad actions, thinking that we are in control. But we (as all other objects in our world) are pulled by this universal force; we (as complex carbon entities) are constantly tugged toward the center.

Since we are alive, we are very efficient at grasping and absorbing the consequences of this force—energy. By continually destroying, we constantly increase entropy, but in so doing attain greater and greater life. We continue to evolve, and in the future our inventions (our machines) will become even more alive than us.

What we consider free will, choice, is predestined. The future exists. We constantly discover it. If we were fourth-dimensional beings, we could actually *see* it. There would be no time, just distinct three-dimensional universes aligned one next to the other. However, we can only comprehend those three dimensions, so we must live each minimally different moment thinking we are moving in time.

## The Charade Of Self-Awareness

*Answer:* I think you understand. Free will is a charade, a game we play in our minds as we decipher the actions we have taken. All actions occur to maximize energy. We have no choice. The universe is so designed. We are self-aware, but we are no different from an inanimate comet hurtling toward the Sun. Our actions are going to happen. In a fourth-dimensional sense, they have already. We are merely discoverers of our own future, one that has existed and will continue to exist for all time.

## Religion

*Question:* But before, when you were discussing math, you called it the logic, the cause and effect, of the universe. You stated that true religion tries to understand this logic; that true religion, not fundamentalism, is the desire to comprehend the workings of God—how the force, how God, is manifested in all. True religion, therefore, incorporates the interpretation of revered books, not just their literal writings. But if what you are now saying is valid, if there is no free will, are not the fundamentalists' allegations correct (that all is willed directly by God, that there is no intermediary force, no cause and effect)?

*Answer:* No, what I said before still holds. The universe has a design—an underlying force guides all. There are causes and effects. We are constantly searching them out. We are extremely efficient; our societies are very sophisticated. We see and interpret their many, many nuances. We exploit these capabilities all the time. Our great ability to learn (to acquire and accumulate knowledge) is why we are so successful.

Our minds, being so complex, allow us to continually monitor our own actions. Therefore, we think we are in charge of what we do; but actually we are merely aware *of what we have done.* We assume we are doing things of our own volition—of free will; in actuality, we do things because we must. It is really God's plan.

## God's Plan

But God's plan is much more complicated than one can describe in a book. A holy book is but an outline; to be deciphered and employed along with the senses given to us. We must use these writings in their truer meaning, not as literal facts but as guides, as schematics of the world. We must use math—logic—also as a guide. Our great strength as thinking, complex living beings is our ability to constantly incorporate all that is about us. This allows us to prosper, to evolve.

So, although all is foreordained, it is brought about by our own actions. We, therefore, are but the conduit for what occurs. We consider these happenings the product of choice, but our undertakings, like all things, are the product of necessity. We rationalize them, we believe in free will, but all are predetermined.

We are simply discoverers of our own destiny. What will occur has already happened, but in a dimension closed to us. We can only enter it as it unfolds. We go hither and yon; our lives travel a zig-zag path. We seek and find, then slip and fall; we grow old and curse our fate. We continuously discover what has already been set out for us; but as we cannot have prior knowledge, we consider our elaborate actions our own. However, we are merely the most advanced performer in carbon's repertoire. Our function is to further the drama, the journey beyond carbon to inorganic forms.   Therefore, our actions are the most efficient (the most sophisticated) in acquiring what is essential—ever-increasing energy, ever-increasing life. Our complexity has become so great that it allows us to ponder what we do; but we do what we must. We think we have free will, but our deeds are predetermined.

**Question:** I guess what you are saying is that the future is set but we cannot know it and therefore think that we can change it. Similarly, our actions occur, they are preordained, but we are unaware, so we seem to have free will. It is really a game that our minds play to keep us interested in our pursuits. They are set, but we cannot know them, so we happily admit to freedom.

**Answer:** Yes, in a nutshell, that is what I think.

**Question:** Well, if the future is fixed, but unknowable, do you still entertain some thoughts of how it may possibly unfold?

# 48

# THE FUTURE

THE UNDERLYING FORCE OF THE UNIVERSE, *gravity*, *leads to increasing efficiency and growth, to evolution.We see this in society's beneficence. Computers, copies of human thought, will continue to improve and finally outperform us. However, for the foreseeable future this greater and greater efficiency, allowing for ever-increasing acquisition, should be of benefit to humans.*

*Health for all should improve as our knowledge of the microbial world increases. More food, better sanitation, and newer sources of energy are attainable as we become ever more sophisticated in our pursuits. Our planet will house more inhabitants and in better conditions in the future.We must continuously strive toward our goals; we can never take them for granted, and what is currently happening in Europe—with the potential dissolution of its Union—should be a lesson for all.*

*One must truly believe in an underlying force that controls our world. I choose to believe in God, the essence of all. Others may disagree, but even avowed atheists need to find sources, or underlying causes. As a scientist, using a higher dimension allows for a more complete understanding—it allows for a belief in God.*

## Efficiency And Growth

*Answer:* No one can predict the future, and it is foolish even to try; however, a broad perspective can be given. It must be based on that underlying principle that we have spoken of so often before, that force that holds us together and keeps us evolving.

Let us assume that there is no devastating war or similar calamity in the next 100 years. If we live in relative peace, then the ongoing forces of more efficient energy procurement will continue as they do today, except at what will seem an ever-increasing pace. Efficiency builds upon efficiency.

In computers, chips will continue to become smaller and carry more information. With ever-increasing computing power, we will be able to converse, and fully interact, with our machines. Much smaller ones may be implanted within our bodies, allowing us significant additional capabilities.

With computers, then, we will continue to increase our data-storage and retrieval ability. We will, thus, become more and more efficient at our primary task: finding and utilizing energy. In 100 years, there is no telling how sophisticated these computers will become, for, by that time, they will probably have surpassed us as a life form. We will be involved in building and maintaining them, but they will be seeking energy for their own purposes as much as for ours.

However, there should be no reason to fear this growth. It will free us, allowing the enjoyment of more varied pursuits. Of course, the quest for leisure for many of us will have become a full-time job. Work will change from manufacturing, as much will be performed by robots, to service. People will be employed in helping others to enjoy this bountiful existence. Thus, a real growth industry should be the recreation field.

In the health arena, with the ever-improving knowledge of genetics, with the ability to design and build artificial life forms, bacteria and other simple organisms will be used to perform many worthwhile tasks. Refuse disposal is a major factor limiting our growth. This problem should be controllable with new, artificially designed bacteria for the solid and liquid wastes that we generate.

New sources of energy and food will certainly be developed using computers and genetic engineering. Superconductors (just starting to be employed) should lead to a fundamental change in the generation and transmission of electricity. Smaller, more efficient motors, much more powerful magnets, and ever more potent computers will come from this. Our planet, in the future, will be able to support more people, and in better health, than it does today.

Transportation will radically change. Anti-gravity devices will be designed. Some think they already have. Roads and bridges will become a thing of the past. Space flight to the edges of our solar system, and beyond, will become a reality. We will easily live and move in the three physical dimensions of our world.

Illnesses that we currently suffer from (like the common cold) will be forgotten in the future, just as is the plague today. Our bodies will be cleaner, for we will be better able to differentiate between helpful and harmful microorganisms. We will not give conveyance to the viruses and bacteria that we so freely carry today, and will not spread disease as we do now.

*Question:* Do you think that increased knowledge and better sanitation alone will be the most important determinants in our future physical health?

*Answer:* Yes, simply knowing what is good or bad for us will allow us to escape many of today's woes. Certainly heart and other vascular "diseases" are problems caused by inadequate knowledge. Infections, and probably a good many cancers, are in many ways also sanitation problems.

As our knowledge expands, and as our ability to safely dispose of wastes increases, we should make great strides toward better health. The most important reasons for improved health in today's world, as compared to the past, are our better standards of sanitation and our increased understanding of the microbial world. As these aspects continue to develop and improve, so will our health.

*Question:* You see the future in a very positive light. You see it as a place where energy will be more efficiently exploited, freeing us to pursue what we wish. Will this not place a great burden on our psyches? Will we not be lost in this overabundance of free time?

*Answer:* I hope not. First of all we have no choice. Only a great unforeseen cataclysm will delay this process. Secondly, if we understand what free time really means, if we truly plan and pursue recreation properly, we should enjoy our lives to a greater extent. Our lives will certainly be less physically painful in the future.

## Euro-Sclerosis

*Question:* You sound very positive, but what about all the problems you discussed concerning our European neighbors? Didn't you say that, unless those countries regain their religious or competitive zeal, they could be inundated by the newly energized and assertive Muslim tide?

*Answer:* Yes, I still think that such an outcome may occur. Much of Europe is borderline socialist, hewing to Marx's notion of "from each according to one's ability to each according to one's need." However, although such a statement sounds noble, the real problem is in defining *need*. Need frequently becomes *desire,* and desire can be infinite; it rarely is satisfied.

Marx's concept only succeeds when one accepts responsibility, when one realizes the true cost of, and is no longer unlimited in, desire. People must believe in a greater essence; there must be a goal beyond mere personal wants. Beliefs based on God have always been the bastions of society. Europe has loosened those anchors and is now adrift. We see a quest for a shared belief in a European Union; but given many different societies, with opposing ideals and prior animosities, such a belief without God is, at best, ephemeral.

Thus, although I feel the future for many will be very favorable we have to continually work toward maintaining our freedom and values.

## One's Belief

*Question:* You have spoken about many things, always finding a primary force. Do you truly believe that our universe has such a common thread?

***Answer:*** You are asking me if I believe in God. I most certainly do. I cannot see how things could exist without such an entity. God's essence is beyond our understanding, but God's actions are somewhat decipherable. The same force that holds things together allows for growth and change.

You and I are lucky to live in perhaps the greatest society ever established. Remember, our founding fathers believed deeply in God. Our Declaration of Independence explicitly states that our rights of life, liberty, and the pursuit of happiness derive directly from our Creator— that they are unalienable.

Newton was profoundly religious; he sincerely felt that God was the essential force guiding all things. Einstein, although not orthodox in his religious beliefs, was also deeply spiritual. His quest was always to find that underlying principle that made sense of the universe.

In this book I have tried to find a glimmer of that encompassing, all-controlling force. To me, it exists as a higher dimension. To me that higher dimension, as it is the basis of all that we can know, is my concept of God. Others may choose to disagree, but even nonbelievers, even avowed atheists, still must believe in primary forces as causes for existence.

I feel it is important for us to try to understand this underlying force. It is important for us to see how we (as individuals and societies), with the curse and benefit of free time, have evolved. If I have helped you to think, to ponder about what before seemed fixed and obvious, then I have accomplished what I have set out to do.

CPSIA information can be obtained
at www.ICGtesting.com
Printed in the USA
FFOW02n0313060115
10070FF